AF899312

An Introduction to
Applied Sport Analytics

An Introduction to
Applied Sport Analytics

JON NACHTIGAL
Assistant Professor and Chair, Sport Business Management
Florida Southern College, Lakeland, FL

DANIEL KRYWARUCZENKO
Sports Science Department Chair
Dallas Christian College, Dallas, TX

ELSEVIER

Academic Press is an imprint of Elsevier
125 London Wall, London EC2Y 5AS, United Kingdom
525 B Street, Suite 1650, San Diego, CA 92101, United States
50 Hampshire Street, 5th Floor, Cambridge, MA 02139, United States

Copyright © 2027 by Elsevier Inc. All rights are reserved, including those for text and data mining, AI training, and similar technologies.

For accessibility purposes, images in electronic versions of this book are accompanied by alt text descriptions independently provided by Elsevier, generated with the aid of AI. For more information, see https://www.elsevier.com/about/accessibility or Radarweg 29, 1043 NX Amsterdam, Netherlands.

Books and Journals published by Elsevier comply with applicable product safety requirements. For any product safety concerns or queries, please contact our authorised representative, Elsevier B.V., at productsafety@elsevier.com or Radarweg 29, 1043 NX Amsterdam, Netherlands.

Publisher's note: Elsevier takes a neutral position with respect to territorial disputes or jurisdictional claims in its published content, including in maps and institutional affiliations.

No part of this publication may be reproduced or transmitted in any form or by any means, electronic or mechanical, including photocopying, recording, or any information storage and retrieval system, without permission in writing from the publisher. Details on how to seek permission, further information about the Publisher's permissions policies and our arrangements with organizations such as the Copyright Clearance Center and the Copyright Licensing Agency, can be found at our website: www.elsevier.com/permissions.

This book and the individual contributions contained in it are protected under copyright by the Publisher (other than as may be noted herein).

Notices

Knowledge and best practice in this field are constantly changing. As new research and experience broaden our understanding, changes in research methods, professional practices, or medical treatment may become necessary.

Practitioners and researchers must always rely on their own experience and knowledge in evaluating and using any information, methods, compounds, or experiments described herein. In using such information or methods they should be mindful of their own safety and the safety of others, including parties for whom they have a professional responsibility.

To the fullest extent of the law, neither the Publisher nor the authors, contributors, or editors, assume any liability for any injury and/or damage to persons or property as a matter of products liability, negligence or otherwise, or from any use or operation of any methods, products, instructions, or ideas contained in the material herein.

ISBN: 978-0-443-49068-2

For Information on all Academic Press publications
visit our website at https://www.elsevier.com/books-and-journals

Publisher: Peter Linsley
Acquisitions Editor: Stephanie Xu
Editorial Project Manager: Vicky Sharma
Project Manager: Prasanna Kalyanaraman
Production Project Manager: Erragounta Saibabu Rao
Designer: Bakyalakshmi S

Typeset by MPS Limited, Chennai, India

ELSEVIER Book Aid International
Working together to grow libraries in developing countries
www.elsevier.com • www.bookaid.org

Last digit is the print number: 9 8 7 6 5 4 3 2 1

CONTENTS

Section IV: Tools of the Trade

Section V: Sharing Your Insights

For additional materials related to the topics in the book, visit the companion site https://www.elsevier.com/books-and-journals/book-companion/9780443490682

For qualified Instructors, additional materials are available in the Instructor site https://educate.elsevier.com/9780443490682

ACKNOWLEDGMENT

My deepest gratitude to my wife, Melissa, whose encouragement and belief in me sustained this project. Her support helped to bring this book to life.

I am grateful to my brother, Jim, for his thoughtful feedback. His insights strengthened this book.

A thank you also goes to Betsy Hipple, who gave me my first position in college softball and welcomed my ideas on using analytics to gain a competitive advantage.

—Jon Nachtigal

I would like to express my deepest gratitude to my wife, Henrit, for her unwavering support, encouragement, and understanding throughout the creation of this textbook. Her patience and love have been a constant source of strength and inspiration during this journey.

—Daniel Krywaruczenko

SECTION I

Laying the Groundwork

This section introduces the history of sport analytics. It begins with a look at how analytics has grown over time and changed the way we understand sports. Then, it explores one of the most important early models, showing how math can help explain team success. These chapters set the stage for everything that comes next by helping you see how numbers tell the story of the game.

CHAPTER 1

The Evolution of Sport Analytics

Key Takeaways

- Sport analytics is the application of objective knowledge to sport, both for the playing field and for the business of sport.
- Baseball served as the foundation for many of the methods and models that shaped modern analytics.
- Advances in technology and the growth of business applications have made analytics central to decisions across the field of sport.

The subject is baseball. The numbers bear a relationship to that subject and to us which is much like the relationship of tools to a machine and to the mechanic who uses them. The mechanic does not begin with a monkey wrench; basically, he is not even interested in the damn monkey wrench. All he wants from the monkey wrench is that it do its job and not give him any trouble. He begins with the machine, with the things which he sees and hears there, and from those he forms an idea – a thesis – about what must be happening in that machine. The tools are a way of taking the thing apart so he can see if he was right or if he needs to try something else.

– Bill James (James, 1982)

From uncovering hidden trends to predicting outcomes, sport analytics has revolutionized our understanding of sport. Today, sport analytics provides teams, coaches, agents, and fans with insights into the game that were either previously unidentified or misunderstood. Sport analytics can be defined as the application of objective, data-driven insights to sport, both on the playing field and in the business of sport. Its principles, however, were slow to catch on.

Baseball has played an important role in the development of sport analytics. In 1859, just before the US Civil War, Henry Chadwick invented the box score to record important statistics from a game (National Baseball Hall of Fame and Museum, n.d.). Chadwick, who has been called the "Father of Baseball," also

An Introduction to Applied Sport Analytics **https://doi.org/10.1016/B978-0-443-49068-2.00001-X**
© 2027 Elsevier Inc. All rights are reserved, including those for text and data mining, AI training, and similar technologies.

contributed the statistic batting average, which has long been questioned as an effective measure of a hitter's ability (Schwarz, 2004). In its original form, batting average used a formula of runs (rather than hits) divided by at bats (Baseball Reference, n.d.).

Henry Chadwick, inventor of the box score and an early developer of baseball statistics.

Over 50 years later, a frustrated F.C. Lane, editor of *Baseball Magazine*, recognized the inadequacies of existing baseball statistics. As an argument against batting average, in 1916, Lane wrote, "Would a system that placed nickels, dimes, quarters and fifty cent pieces on the same basis be much of a system whereby to compute a man's financial resources? Anyone who offered such a system would deserve to be examined as to his mental condition. And yet it is precisely such a loose, inaccurate system which obtains in baseball and lies at the root of the most popular branch of baseball statistics" (Lane, 1916a).

Lane set to work developing a more accurate method for evaluating hitters than batting average by recording 1000 hits and tracking their outcomes. As a result, he determined weights for singles, doubles, triples, and home runs, assigning values of 0.30, 0.60, 0.90, and 1.15, respectively (Lane, 1916b). These values corresponded with the number of runs each event generated on average. Lane wrote, "Let it be hoped that 1916, the dawn of a new day in baseball affairs, will witness as well the dawn of a new day in the outworn method of keeping batting averages" (Lane, 1916a). Although it would take almost a century for his work to catch on, Lane laid the groundwork for today's models that use linear weights (see Chapter 7) for baseball events to better determine a player's true ability.

In 1947, Alan Roth became baseball's first full-time statistician when he was hired by the Brooklyn Dodgers. Roth provided innovative analysis that had not been previously explored. In his work for the Brooklyn and Los Angeles Dodgers, Roth analyzed:

1. Bunting situations
2. Park effects, which measure how the dimensions of the stadium affect player performance
3. Platoon splits, which evaluate the impact of a batter's and pitcher's handedness on the game
4. Player performance during day and night games
5. Spray charts to evaluate an opponent's tendencies for where they would hit the ball
6. Team winning percentage when at home versus away

Roth's analytical prowess was considerably ahead of his time, which benefited the Dodgers. After a change in Dodgers ownership, Roth moved to the press box, eventually contributing his statistics to the work of legendary baseball announcer Vin Scully (Schwarz, 2004).

Dr. George Lindsay, a Canadian nuclear physicist, made significant contributions to the field of baseball analytics in the 1960s. In addition to his work as Chief of the Operational Research and Analysis with the Canadian Department of National Defence, Lindsay also applied his research skills to baseball, publishing three papers on the sport. Lindsay was the first to measure run expectancy from the 24 base and out states (see Chapter 7 for additional information on run expectancy), he evaluated various strategies used in the game such as intentional walks and stolen bases, and he was the first to identify how win probability changes throughout a game based on the score.

The figure responsible for bringing sport analytics into the mainstream is Bill James. Despite his long-term success, James' beginnings were humble. In 1977, James self-published his first *Baseball Abstract* and placed an ad in the back of *The Sporting News* to promote his work. His research was performed by hand, as computers were in their infancy at the time. And though it took a few years, eventually he found an audience of like-minded individuals (Schwarz, 2004).

Unlike today, where we are awash in data, James struggled with a lack of baseball data. There's a saying that research is as good as its data. Unfortunately for James, baseball data was unavailable to fans since it was controlled by Elias Sports Bureau and Major League Baseball. As described by Michael Lewis in his book *Moneyball*:

> *The customer wanted stats and Major League Baseball did its best not to give them to him. The people inside Major League Baseball were, if anything, hostile to the people outside Major League Baseball who wished to study the game. That struck James, who by now had perfected the art of sounding like a sane man in an insane world, as mad.*[7]

Frustrated by his failure to acquire data from Elias and Major League Baseball, James enlisted an army of baseball fans who began documenting play-by-play information for his Project Scoresheet. This was critically important because, for the first time, James enabled fans to have open access to baseball data (Schwarz, 2004).

Bill James, founder of sabermetrics, who helped shape modern baseball analytics.

James would question long-held beliefs about baseball, insisting on a scientific approach to better understanding the game. He was also prolific, introducing a variety of statistics that had never been seen before. For example, his Pythagorean Theorem of Baseball formula, introduced in 1983, uses a team's runs scored and runs allowed to accurately predict winning percentage (James, 1983). Since its introduction, this formula, with varying optimal exponents, has been used for a variety of sports to estimate winning percentages.

By 1984, just 7 years after self-publishing his first baseball abstract, *The Bill James Baseball Abstract* had risen to number four on the *New York Times* bestseller list. As a result, sport analytics began to enter the mainstream (Schwarz, 2004).

Another watershed moment for sports analytics came with the publication of Michael Lewis' book *Moneyball* in 2003, which chronicled the Oakland Athletics' innovative use of analytics under General Manager Billy Beane. *Moneyball* described a new era where data-driven decision-making became central to baseball strategy and player evaluation. *Moneyball* helped popularize sport analytics, with teams and businesses alike eventually embracing a scientific approach to analytics, leading to its widespread adoption and integration into most aspects of modern life.

Many other figures have helped us to better understand sports through analytics, such as Pete Palmer and John Thorn, who published analyses of football and baseball in the 1980s; Jon Dewan, Jim Albert, Vörös McCracken, John Walsh, and Max Marchi, who have written extensively on the subject, to name a few; and then Tom Tango, Mitchel Lichtman, and Andrew Dolphin, who in 2006 published their seminal book on baseball analytics, *Playing the Percentages in Professional Baseball.* There is a lighthearted look at applying analytics to an independent professional baseball team in *The Only Rule Is It Has to Work* by Ben Lindbergh and Sam Miller. Then there's Keith Law, who looked at advances in baseball analysis in his book *Smart Baseball.* For researching the National Football League (NFL), there are *Football Analytics with Python & R* by Eric Eager and Richard Erickson and *Introduction to NFL Analytics with R* by Bradley Congelio. For basketball, there's *Basketball on Paper* by Dean Oliver. For whatever sport you are interested in, there's probably a book available to learn more about it from an analytics perspective.

Despite significant progress made in the field of sports analytics, there remains a notable gender gap in the representation of female athletes. Allison McCann, in an article on *FiveThirtyEight.com*, described this disparity. In her article, McCann wrote, "And while you can easily look up all 14,260,129 at-bats in the history of Major League Baseball, I have no idea how many at-bats were taken during the five years of the Women's Professional Softball League. That league folded—along with any of the data it recorded, presumably." McCann also pointed to the lack of data for women's professional golf in comparison to men's professional golf (McCann, 2015).

The gender gap also exists in academic journals, where female athletes are analyzed less than their male counterparts (Brookshire, 2016). As a result, women remain underrepresented in analytical studies. This underrepresentation limits the depth and breadth of available analytics, leaving fewer datasets and models to inform our understanding of women's athletic performance.

As an example of how far sport analytics has developed, consider Catapult Sports, a leader in wearable tracking technology that is used by over 4600 elite teams (Catapult, n.d.). Its devices use GPS to measure how athletes accelerate, decelerate, change directions, and jump (Mackay et al. 2025). This technology illustrates how the idea of collecting and analyzing sport data, which was once done with paper and pencil, now happens in real time.

Thanks to the pioneers in the field and advances in technology, sport analytics is now a tool that teams can leverage to gain a competitive advantage. Today, however, the challenge has shifted from too little data to what, at times, seems like an overwhelming volume of data. As an example, Major League Baseball's Statcast system generates roughly 7 terabytes of data per game (Krohn, 2025).

Regarding the business of sport, most jobs in analytics focus on improving efficiencies for the sport organization off the playing field rather than on it. Analytics has become a cornerstone for business decision-making and is a powerful tool that can effectively drive organizational transformation. Sport organizations are using business analytics to boost customer acquisition and retention, improve fan engagement, solve advertising and sponsorship challenges, and provide insights into marketing strategies to improve the organization's bottom line. By leveraging the power of data, sport organizations are making more informed business decisions, improving their efficiency and profitability.

SUMMARY

This section traces the evolution of sport analytics from its earliest roots in baseball through its emergence as a transformative force in sport. Pioneers such as Henry Chadwick, F.C. Lane, Alan Roth, George Lindsay, and especially Bill James laid the foundation for modern analytics. Despite these advancements, gaps remain, as in the limited availability of women's sports data. Even so, sport analytics has become an essential asset for organizations seeking both competitive success and strategic business growth.

BEYOND THE BOX SCORE

1. Bill James compares sport analytics to a mechanic using tools. What does this metaphor suggest about the role of statistics in understanding sport performance? How might this influence your approach to using data in sport analytics?
2. F.C. Lane and others challenged long-standing statistics such as batting average. Can you think of a sport today where a traditional metric might still dominate, even if it's flawed? Why do you think it persists?
3. What common thread connects early pioneers such as Chadwick, Lane, Roth, and Lindsay to modern sport analysts? How can these historical examples inspire your own use of data in either sport performance or sport business?
4. Reflect on the idea that "the problem today isn't too little data, it's too much." What skills or strategies do you think are most important for an analyst trying to navigate modern data overload in sports?
5. Allison McCann discusses the lack of data for women's sports. How might this imbalance affect the growth, visibility, or investment in women's sports? What role can sport analytics students or professionals play in addressing this gap?

AI DISCLOSURE

During the preparation of this work, the author(s) used ChatGPT (OpenAI GPT-5) in order to provide editing and improve the readability of the document. After using this tool/service, the author(s) reviewed and edited the content as needed and take(s) full responsibility for the content of the publication.

REFERENCES

Baseball Reference, n.d. Henry Chadwick. https://www.baseball-reference.com/bullpen/Henry_Chadwick.

Brookshire, B., 2016. Women in sports are often underrepresented in science: Studies of competitive sports and exercise are still dominated by men. *Science News*. https://www.sciencenews.org/blog/scicurious/women-sports-are-often-underrepresented-science.

Catapult, n.d. About Catapult. https://www.catapult.com/company/about-catapult (accessed 26.09.25).

James, B., 1982. Bill James Baseball Abstract. Lawrence: Self-published.

James, B., 1983. The Pythagorean Theorem of Baseball. *Bill James Baseball Abstract*.

Krohn, J., 2025. How AI is transforming baseball (with lessons for all of us). https://www.jonkrohn.com/posts/2025/3/28/how-ai-is-transforming-baseball-with-lessons-for-all-of-us.

Lane, F.C. (1916a). Analysis and new metrics for baseball. Baseball Magazine.

Lane, F.C. (1916b). Methodology for assigning run values to hits. Baseball Magazine.

Lewis, M., 2003. Moneyball: The Art of Winning an Unfair Game. W. W. Norton.

Mackay, L., Sawczuk, T., Jones, B., Darrall-Jones, J., Clark, A., Whitehead, S., 2025. The reliability of a commonly used (Catapult™ Vector S7) microtechnology unit to detect movement characteristics used in court-based sports. Journal of Sports Sciences, 43(6), 555–564. https://doi.org/10.1080/02640414.2025.2468585.

McCann, A., 2015. Hey, Nate: There is no "rich data" in women's sports. FiveThirtyEight. Available at: https://fivethirtyeight.com/features/hey-nate-there-is-no-rich-data-in-womens-sports/ (accessed 29.06.26).

National Baseball Hall of Fame and Museum, n.d. Henry Chadwick. https://baseballhall.org/hall-of-famers/chadwick-henry (accessed 04.07.25).

Schwarz, A., 2004. The Numbers Game. Thomas Dunne Books, New York.

CHAPTER 2

The Pythagorean Theorem of Sports

Key Takeaways

- Bill James' Pythagorean Theorem of Baseball demonstrated the relationship between runs scored, runs allowed, and winning percentage.
- The formula also provides an opportunity to discuss randomness in sport, as the role of chance plays a part in team wins and losses.
- The formula can be adapted to other team sports, making it a foundational tool in sport analytics.

Baseball, it is said, is only a game. True. And the Grand Canyon is only a hole in Arizona. Not all holes, or games, are created equal.

– George Will (Will and Work, 1990)

Bill James called his approach to analyzing baseball "sabermetrics," in honor of the Society of American Baseball Research (SABR) (James, 1982). James described sabermetrics as "the search for objective knowledge about baseball" (*What is sabermetrics?*, n.d.). It's easy to see today how sabermetrics and this approach of applying science to baseball research would evolve into what is referred to as analytics. One of James' key contributions to sport analytics was his Pythagorean Theorem. James recognized that a relationship exists between the number of runs teams score during a season, the number of runs they allow, and their winning percentages. This relationship can be demonstrated in the final standings in baseball's American League East division in 1979 (See Fig. 2.1 below).

An Introduction to Applied Sport Analytics **https://doi.org/10.1016/B978-0-443-49068-2.00002-1**
© 2027 Elsevier Inc. All rights are reserved, including those for text and data mining, AI training, and similar technologies.

AL EAST	W	L	PCT	RS	RA	DIFF
Baltimore Orioles	102	57	.642	757	582	+175
Milwaukee Brewers	95	66	.590	807	722	+85
Boston Red Sox	91	69	.569	841	711	+130
New York Yankees	89	71	.556	734	672	+62
Detroit Tigers	85	76	.528	770	738	+32
Cleveland Indians	81	80	.503	760	805	-45
Toronto Blue Jays	53	109	.327	613	862	-249

Figure 2.1 1979 AL East standings with run differential.

As you can see in the 1979 standings, the teams largely finished in order of their run differential (DIFF), which is runs scored (RS) minus runs allowed (RA). Except for the Milwaukee Brewers finishing ahead of the Boston Red Sox, run differential correlated with the finish of every team in the division.

In 1980, James first mentioned his formula for use in estimating a team's chances of winning (James, 1980). The formula, known as the Pythagorean Theorem of Baseball, honored the ancient Greek mathematician and philosopher Pythagoras, whose Pythagorean Theorem is used in measuring right triangles. The Pythagorean Theorem of Pythagoras found that the square of the length of the hypotenuse (the side opposite the right angle) is equal to the sum of the squares of the lengths of the other two sides. Mathematically, this theorem is expressed as:

$$a^2 + b^2 = c^2$$

James referenced Pythagoras with his Pythagorean Theorem of Baseball due to the similarity in the formulas. Instead of the three sides of a triangle, James' formula showed that there's a relationship between how many runs or points teams score over the course of a season (a^2), how many runs or points teams allow in a season (b^2), and how many games they win (c^2). James' initial formula for analyzing professional baseball used an exponent of 2 and looked like:

$$\text{Runs Scored}^2 \,/\, (\text{Runs Scored}^2 + \text{Runs Allowed}^2) = \text{Winning Percentage}$$

By adjusting the exponent in the formula from 2 to around 1.83, a later version of the Pythagorean Theorem of Baseball became even more accurate (*Pythagorean Theorem of Baseball*, n.d.). As shown in the following image, here's the predicted won and loss (X-W/L) record of each team based on the Pythagorean Theorem of Baseball (MLB standings and records: Regular season, 1979) (Fig. 2.2).

AL EAST	W	L	PCT	RS	RA	DIFF	X-W/L
Baltimore Orioles	102	57	.642	757	582	+175	98-61
Milwaukee Brewers	95	66	.590	807	722	+85	89-72
Boston Red Sox	91	69	.569	841	711	+130	92-68
New York Yankees	89	71	.556	734	672	+62	86-74
Detroit Tigers	85	76	.528	770	738	+32	84-77
Cleveland Indians	81	80	.503	760	805	-45	76-85
Toronto Blue Jays	53	109	.327	613	862	-249	57-105

Figure 2.2 1979 AL East standings with predicted won and loss (X-W/L).

Looking again at the AL East in 1979, the column for expected wins and losses (X-W/L) uses the Pythagorean Theorem of Baseball to predict what each team's record would have been without randomness or luck. While the Baltimore Orioles won four more games than would have been expected according to the formula, they still would have been expected to easily win the division. The Brewers were also the luckiest team that season, winning six more games than was expected. Although the Toronto Blue Jays finished last that season, they weren't as bad as their record. According to the formula, the Jays should have won four more games.

In addition to its use in predicting team winning percentages, the creation of the Pythagorean Theorem of Baseball has allowed analysts and fans to project player impact on winning percentage. In baseball, it's known that about every 10 runs in run differential is worth one win (*WAR explained: Converting runs to wins*, n.d.). Therefore, if a team is considering a player who is projected to save or score 10 runs when compared to the player who is currently on the roster, that player should be worth one more win. This opened the door to quantifying player contributions to wins. James would go on to create additional formulas to assist in the evaluation of players. One such formula, Runs Created, evaluates a player's expected contribution to scoring (and ultimately wins) with the following formula (*Runs created (RC)*, n.d.):

$$\text{Total Bases} \star (\text{Hits} + \text{Walks})/(\text{At Bats} + \text{Walks})$$

THE PYTHAGOREAN THEOREM OF SPORTS

Importantly, James' formula can be applied to sports other than just baseball. The Pythagorean Theorem of Sports formula works for basketball, football, hockey, soccer, and most team sports. For example, in NCAA Division I softball, the formula exhibits remarkable accuracy in predicting team winning percentages, just like in other sports. Using an exponent of 1.55 (the exponent varies by sport), you can see in the following image that the Pythagorean Theorem of Softball accurately predicts team winning percentages. Each dot on the image represents 1 of the 295 teams in NCAA Division I softball in the 2014 season (Fig. 2.3).

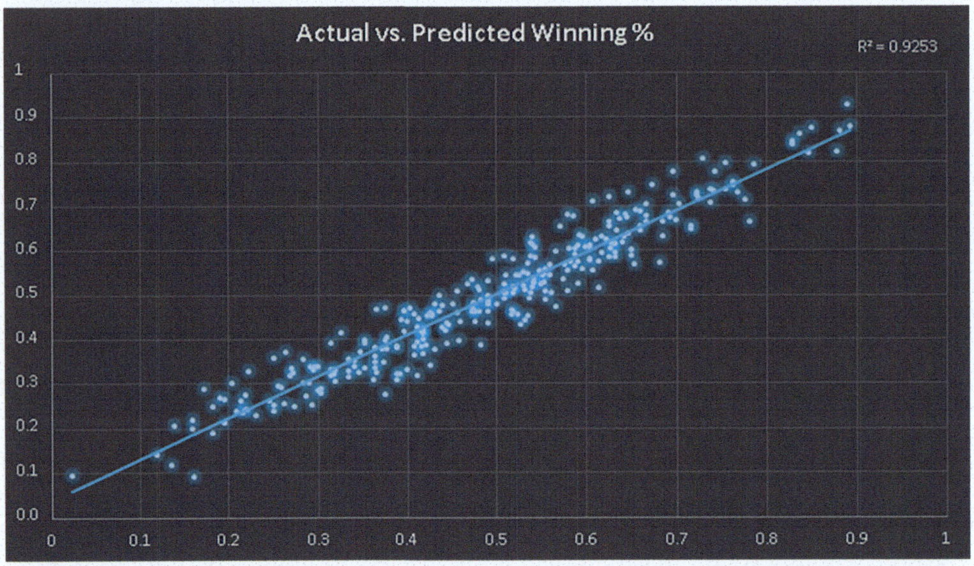

Figure 2.3 Actual versus predicted win percentage for NCAA Division I softball in the 2014 season.

The accuracy of the formula can be demonstrated with how each team's actual and predicted winning percentage is plotted near the trend line. In addition, the *R*-squared (R^2), or coefficient of determination, of 0.9253 is incredibly high. The scale for R^2 ranges from 0 to 1.0. An R^2 of 0.9253 means that almost 93% of the variance in the data is explained by the formula and about 7% of the variance is due to unexplained or other effects. This indicates that the formula does a very good job of matching the data and predicting the outcomes (winning percentage) based on the inputs (runs scored and runs allowed).

The power of Bill James' Pythagorean Theorem lies not only in analyzing the past but also in projecting the future. In the next chapter, we will apply the formula to real-world datasets, providing hands-on experience with one of sport analytics' most influential tools.

SUMMARY

This section explains the origins and significance of Bill James' Pythagorean Theorem of Sports, a foundational concept in sport analytics. By linking a team's scoring and defensive performance to its expected win percentage, the formula helps identify team strength and also the role of randomness. Initially developed for baseball, the theorem has since been adapted across multiple sports and leagues. Beyond team performance, the formula can be used by analysts to assess player contributions in terms of wins, laying the groundwork for future evaluation tools.

BEYOND THE BOX SCORE

1. Bill James borrowed the name "Pythagorean Theorem" from geometry to describe his baseball formula. Why do you think drawing from mathematical or scientific concepts helps legitimize ideas in sport analytics?
2. The 1979 American League East standings largely followed the order of run differential. What does this suggest about randomness in sport, and how can formulas like the Pythagorean Theorem help us distinguish between luck and skill?
3. How might the ability to estimate wins based on run differential affect how teams make roster decisions or value players in trades and contracts?
4. James' formula has now been adapted to other sports, each with a different exponent. What challenges might arise when applying a single formula across sports with very different scoring systems and game dynamics?
5. The R^2 value of 0.9253 in the softball example suggests strong predictive power. But even a highly accurate model leaves room for unpredictability. How should coaches, analysts, and teams account for this unpredictability when using data to make decisions?

LOOKING AHEAD

In the next section, you will use Excel to work with data and practice the fundamentals of sport analytics. You'll begin by applying what you learned about the Pythagorean Theorem by using the formula with real data and then move into core tools such as correlation and regression. Together, these fundamentals provide a foundation for evaluating data in sport analytics.

AI DISCLOSURE

During the preparation of this work, the author(s) used ChatGPT (OpenAI GPT-5) in order to provide editing and improve the readability of the document. After using this tool/service, the author(s) reviewed and edited the content as needed and take(s) full responsibility for the content of the publication.

REFERENCES

Baseball-Reference, n.d. Pythagorean Theorem of Baseball, BR Bullpen. Available at: https://www.baseball-reference.com/bullpen/Pythagorean_Theorem_of_Baseball (accessed 03.06.25.).

Baseball-Reference, n.d. WAR explained: Converting runs to wins. Available at: baseball-reference.com (accessed 03.06.25.).

James, B., 1980. 1980 Baseball Abstract. Self-published.

James, B., 1982. Bill James Baseball Abstract, Self-published.

Major League Baseball, n.d. 1979 MLB standings and records: Regular season. Available at: https://www.mlb.com/standings/1979 (accessed 03.06.25.).

MLB Advanced Media, n.d. Runs created (RC). Available at: https://www.mlb.com/glossary/advanced-stats/runs-created (accessed 11.07.25.).

Society for American Baseball Research, n.d. What is sabermetrics? Available at: https://sabr.org/sabermetrics/ (accessed 15.07.26.).

Will, G. F., Work, M. A., 1990. The Craft of Baseball.

SECTION II

Applying the Fundamentals

In this section, you will start using the tools of sport analytics yourself. Each chapter introduces a key method and shows how to work with data using Microsoft Excel. These are the building blocks of data analysis and will help you begin answering real questions with data.

CHAPTER 3

Applying the Pythagorean Theorem

Key Takeaways

- The Pythagorean Theorem demonstrates how scoring differential can be used to calculate expected team wins.
- The formula allows us to compare predicted and actual wins, with visualizations and R^2 values confirming its effectiveness.
- Beyond analyzing past outcomes, the formula can also be used to project future results, offering valuable insight.

I try to take large, general questions that are difficult to resolve and break them down into small, very specific questions that have clear answers.

– Bill James (James, 2015)

While it is one thing to talk about sport analytics, it is quite another challenge to perform the actual analysis. Let's apply what we have learned about the Pythagorean Theorem of Baseball. Using the data from the Major League Baseball season in 1979 that was introduced in Chapter 2, we will apply the formula for Pythagorean Theorem of Baseball to determine the expected winning percentage of teams. Remember, the formula is based on runs scored and runs allowed. Here is an updated version of the formula for the Pythagorean Theorem of Baseball with an exponent of 1.83:

$$\text{Runs Scored}^{1.83}\ \text{Runs Score}^{1.83} + \text{Runs Allowed}^{1.83}$$

Let's use Microsoft Excel to apply the Pythagorean Theorem of Baseball to determine the expected number of wins (X-W) for each team.

1. Download the file "MLB Standings 1979.xlsx" from the book's website.
2. Open the file in Excel. You will see the year-end standings for every team in MLB in 1979 (Fig. 3.1).

An Introduction to Applied Sport Analytics **https://doi.org/10.1016/B978-0-443-49068-2.00003-3**
© 2027 Elsevier Inc. All rights are reserved, including those for text and data mining, AI training, and similar technologies.

	A	B	C	D	E	F	G	H
1	TEAM	W	L	PCT	RS	RA	GP	X-W
2	Baltimore Orioles	102	57	0.642	757	582	159	
3	Pittsburgh Pirates	98	64	0.605	775	643	162	
4	Montreal Expos	95	65	0.594	701	581	160	
5	Milwaukee Brewers	95	66	0.590	807	722	161	
6	Boston Red Sox	91	69	0.569	841	711	160	
7	Cincinnati Reds	90	71	0.559	731	644	161	
8	New York Yankees	89	71	0.556	734	672	160	
9	Houston Astros	89	73	0.549	583	582	162	
10	California Angels	88	74	0.543	866	768	162	
11	St. Louis Cardinals	86	76	0.531	731	693	162	
12	Detroit Tigers	85	76	0.528	770	738	161	
13	Kansas City Royals	85	77	0.525	851	816	162	
14	Philadelphia Phillies	84	78	0.519	683	718	162	
15	Texas Rangers	83	79	0.512	750	698	162	
16	Minnesota Twins	82	80	0.506	764	725	162	
17	Cleveland Indians	81	80	0.503	760	805	161	
18	Chicago Cubs	80	82	0.494	706	707	162	
19	Los Angeles Dodgers	79	83	0.488	739	717	162	
20	Chicago White Sox	73	87	0.456	730	748	160	
21	San Francisco Giants	71	91	0.438	672	751	162	
22	San Diego Padres	68	93	0.422	603	681	161	
23	Seattle Mariners	67	95	0.414	711	820	162	

Figure 3.1 Excel output showing year-end standing of MLB teams in 1979.

3. In cell H2, type the formula for the Pythagorean Theorem of Baseball. The formula shown below is: Runs Scored to an exponent of 1.83, divided by Runs Scored to an exponent of 1.83 plus Runs Allowed to an exponent of 1.83, multiplied by the total number of games played, which differed by team. In Excel, in cell G2 type (Fig. 3.2):

	A	B	C	D	E	F	G	H	I	J	K
1	TEAM	W	L	PCT	RS	RA	GP	X-W			
2	Baltimore Orioles	102	57	0.642	757	582	159	=(E2^1.83/(E2^1.83+F2^1.83))*G2			
3	Pittsburgh Pirates	98	64	0.605	775	643	162				
4	Montreal Expos	95	65	0.594	701	581	160				
5	Milwaukee Brewers	95	66	0.590	807	722	161				
6	Boston Red Sox	91	69	0.569	841	711	160				
7	Cincinnati Reds	90	71	0.559	731	644	161				
8	New York Yankees	89	71	0.556	734	672	160				
9	Houston Astros	89	73	0.549	583	582	162				
10	California Angels	88	74	0.543	866	768	162				
11	St. Louis Cardinals	86	76	0.531	731	693	162				
12	Detroit Tigers	85	76	0.528	770	738	161				
13	Kansas City Royals	85	77	0.525	851	816	162				
14	Philadelphia Phillies	84	78	0.519	683	718	162				
15	Texas Rangers	83	79	0.512	750	698	162				
16	Minnesota Twins	82	80	0.506	764	725	162				
17	Cleveland Indians	81	80	0.503	760	805	161				
18	Chicago Cubs	80	82	0.494	706	707	162				
19	Los Angeles Dodgers	79	83	0.488	739	717	162				
20	Chicago White Sox	73	87	0.456	730	748	160				
21	San Francisco Giants	71	91	0.438	672	751	162				
22	San Diego Padres	68	93	0.422	603	681	161				
23	Seattle Mariners	67	95	0.414	711	820	162				

Figure 3.2 Excel output showing the formula for calculating expected wins (X-W).

$$= (E2^{\wedge}1.83/(E2^{\wedge}1.83 + F2^{\wedge}1.83)) \star G2$$

Press the Enter key. The cell should display the value of 98.26. This is the projected number of wins the Orioles should have won based on their runs scored and runs allowed.

1. Left click in cell G2, then left double click the box in the lower right corner of cell H2 or left click and drag down to cell H8 to populate the entire column. The expected number of wins (X-W) for each team will appear (Fig. 3.3).

	A	B	C	D	E	F	G	H
1	TEAM	W	L	PCT	RS	RA	GP	X-W
2	Baltimore Orioles	102	57	0.642	757	582	159	98.3
3	Pittsburgh Pirates	98	64	0.605	775	643	162	
4	Montreal Expos	95	65	0.594	701	581	160	
5	Milwaukee Brewers	95	66	0.590	807	722	161	
6	Boston Red Sox	91	69	0.569	841	711	160	
7	Cincinnati Reds	90	71	0.559	731	644	161	
8	New York Yankees	89	71	0.556	734	672	160	
9	Houston Astros	89	73	0.549	583	582	162	
10	California Angels	88	74	0.543	866	768	162	

Figure 3.3 Excel output showing the expected wins for the Baltimore Orioles.

Let's see just how accurate our formula is by creating a scatter chart. A scatter chart is a graph that displays individual data points to show the relationship between two variables. To create a scatter chart showing the relationship between the actual number of wins and the expected number of wins, in Excel:

1. Let's compare the columns Wins (W) and expected wins (X-W). First, left-click and drag to highlight the Wins column. Then, hold the CTRL key, and left-click and drag the expected wins column (Fig. 3.4).

	A	B	C	D	E	F	G	H
1	TEAM	W	L	PCT	RS	RA	GP	X-W
2	Baltimore Orioles	102	57	0.642	757	582	159	98.3
3	Pittsburgh Pirates	98	64	0.605	775	643	162	94.7
4	Montreal Expos	95	65	0.594	701	581	160	93.6
5	Milwaukee Brewers	95	66	0.590	807	722	161	88.7
6	Boston Red Sox	91	69	0.569	841	711	160	92.2
7	Cincinnati Reds	90	71	0.559	731	644	161	89.8
8	New York Yankees	89	71	0.556	734	672	160	86.4
9	Houston Astros	89	73	0.549	583	582	162	81.1
10	California Angels	88	74	0.543	866	768	162	89.9
11	St. Louis Cardinals	86	76	0.531	731	693	162	85.0
12	Detroit Tigers	85	76	0.528	770	738	161	83.6
13	Kansas City Royals	85	77	0.525	851	816	162	84.1
14	Philadelphia Phillies	84	78	0.519	683	718	162	77.3
15	Texas Rangers	83	79	0.512	750	698	162	86.3

Figure 3.4 Excel output showing the X-W column populated with calculated values.

2. From the Insert tab, click the drop-down for Scatter Charts and choose Scatter (Fig. 3.5).

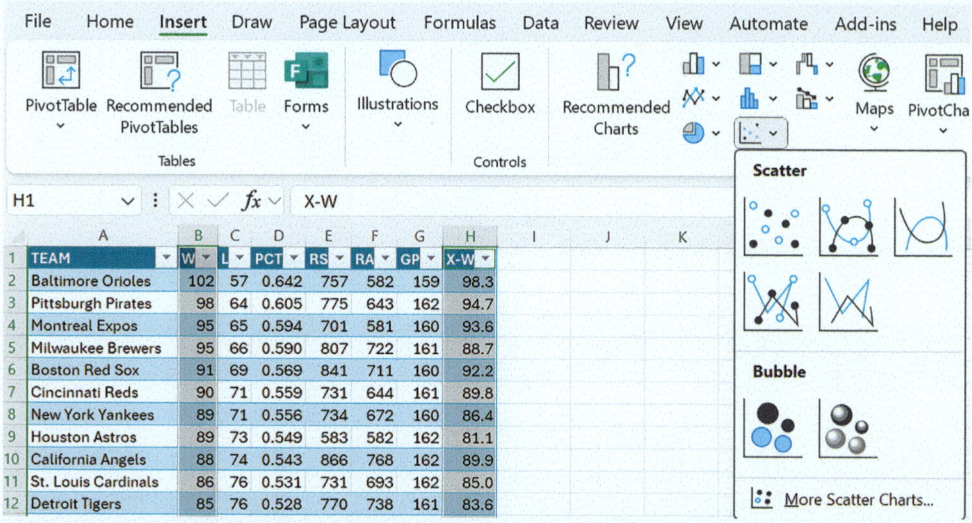

	A	B	C	D	E	F	G	H
1	TEAM	W	L	PCT	RS	RA	GP	X-W
2	Baltimore Orioles	102	57	0.642	757	582	159	98.3
3	Pittsburgh Pirates	98	64	0.605	775	643	162	94.7
4	Montreal Expos	95	65	0.594	701	581	160	93.6
5	Milwaukee Brewers	95	66	0.590	807	722	161	88.7
6	Boston Red Sox	91	69	0.569	841	711	160	92.2
7	Cincinnati Reds	90	71	0.559	731	644	161	89.8
8	New York Yankees	89	71	0.556	734	672	160	86.4
9	Houston Astros	89	73	0.549	583	582	162	81.1
10	California Angels	88	74	0.543	866	768	162	89.9
11	St. Louis Cardinals	86	76	0.531	731	693	162	85.0
12	Detroit Tigers	85	76	0.528	770	738	161	83.6

Figure 3.5 Adding scatter chart to output.

3. A trendline is a straight or curved line that shows the general direction or pattern of data points. To add a trendline, right-click any of the data points on the scatter chart and choose Add Trendline (Fig. 3.6).

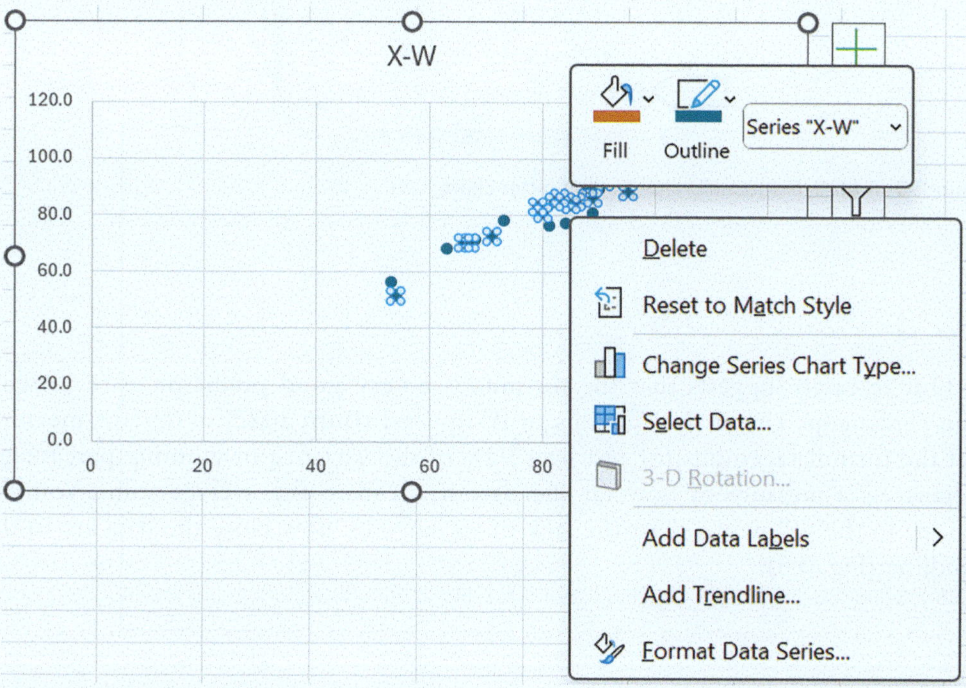

Figure 3.6 Creating a trendline.

4. Place a check in the check box to add the R-squared value to the chart (Fig. 3.7).

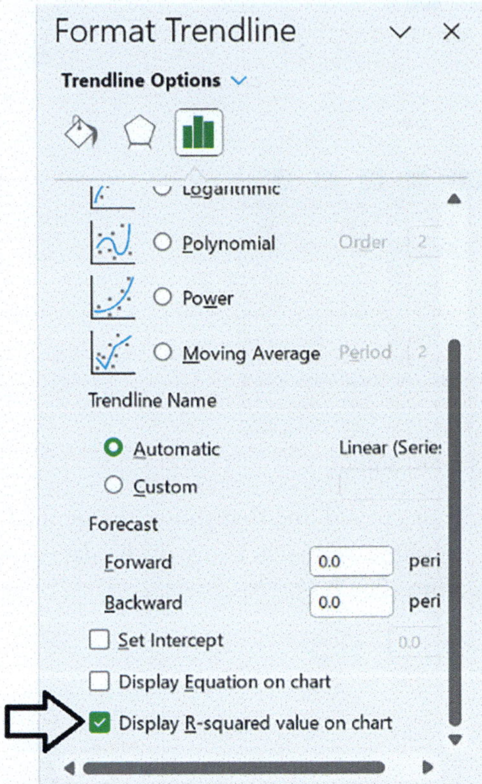

Figure 3.7 Adding R-squared value to the scatter chart.

Our research suggests that the formula is a very good predictor of team winning percentage. The analysis reveals an R-squared (Butt, 2022) of 0.9202, meaning that the formula accounts for just over 92% of the variance in winning percentage. Variance is a measure of how far the data strays from the average value. You can also see in the following chart the lack of variance by how close the data are to the trendline (Fig. 3.8).

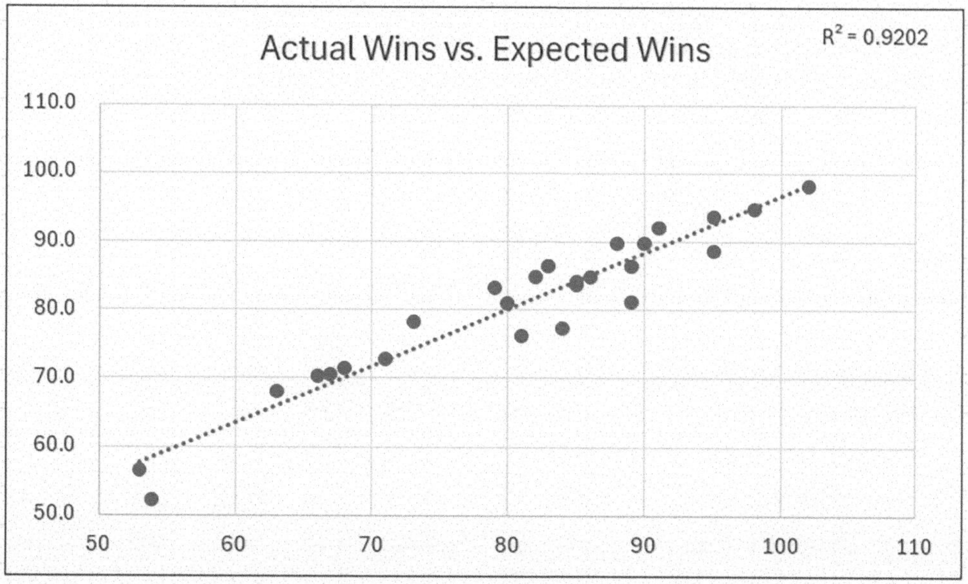

Figure 3.8 Actual wins versus expected wins scatter chart.

The title of the chart can be changed by clicking on the title text. To adjust the axes, left-click an axis, then right-click, and select Format Axis (Fig. 3.9).

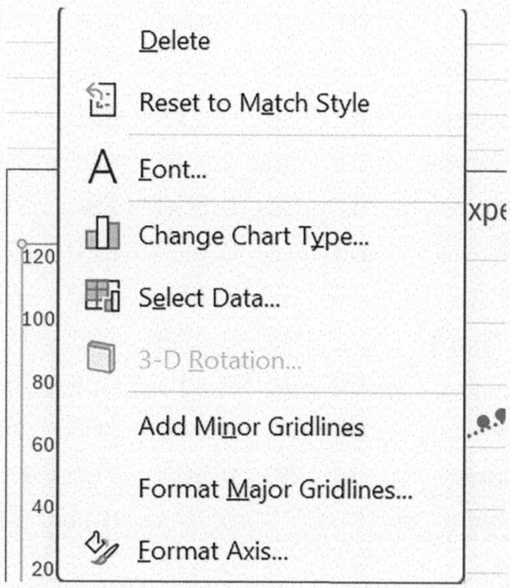

Figure 3.9 How to select Format Axis... in Excel.

PROJECTING WINS

It is one thing to analyze the past; it is entirely different to predict the future. By applying the Pythagorean Theorem of Baseball, we now have a reliable tool that can do just that. This formula allows us to project a team's expected future record with remarkable accuracy.

Let's say MLB is going to expand with two more teams, like they did by adding the Colorado Rockies and the Florida Marlins in 1993. The analytics staff for these two teams, much like is documented in *Moneyball*, has developed a projection for how many runs each team will score and allow over the course of a season. Here are the estimates for both teams:

- Colorado Rockies: 760 runs scored, 965 runs allowed
- Florida Marlins: 580 runs scored, 725 runs allowed

Populate the spreadsheet with the above information in rows 28 and 29. Your spreadsheet should look like this (Fig. 3.10).

	A	B	C	D	E	F	G	H
4	Baltimore Orioles	102	57	0.642	757	582	159	98.3
5	Pittsburgh Pirates	98	64	0.605	775	643	162	94.7
6	Cincinnati Reds	90	71	0.559	731	644	161	89.8
7	New York Yankees	89	71	0.556	734	672	160	86.4
8	San Diego Padres	68	93	0.422	603	681	161	71.6
9	St. Louis Cardinals	86	76	0.531	731	693	162	85.0
10	Texas Rangers	83	79	0.512	750	698	162	86.3
11	New York Mets	63	99	0.389	593	706	162	68.2
12	Chicago Cubs	80	82	0.494	706	707	162	80.9
13	Boston Red Sox	91	69	0.569	841	711	160	92.2
14	Los Angeles Dodgers	79	83	0.488	739	717	162	83.2
15	Philadelphia Phillies	84	78	0.519	683	718	162	77.3
16	Milwaukee Brewers	95	66	0.590	807	722	161	88.7
17	Minnesota Twins	82	80	0.506	764	725	162	84.9
18	Detroit Tigers	85	76	0.528	770	738	161	83.6
19	Chicago White Sox	73	87	0.456	730	748	160	78.2
20	San Francisco Giants	71	91	0.438	672	751	162	72.8
21	Atlanta Braves	66	94	0.413	669	763	160	70.4
22	California Angels	88	74	0.543	866	768	162	89.9
23	Cleveland Indians	81	80	0.503	760	805	161	76.3
24	Kansas City Royals	85	77	0.525	851	816	162	84.1
25	Seattle Mariners	67	95	0.414	711	820	162	70.5
26	Oakland Athletics	54	108	0.333	573	860	162	52.2
27	Toronto Blue Jays	53	109	0.327	613	862	162	56.5
28	Colorado Rockies				760	965		
29	Florida Marlins				580	725		

Figure 3.10 Spreadsheet with more expected wins filled into the respective Excel output.

We also know that both teams should play the full schedule of 162 games. Once you populate this information in column G, you see that in column H the formula gives us how many games each team is expected to win based on the information we provided. If column H doesn't automatically populate, left-click and drag cell H27 down to cell H29. The spreadsheet for these two teams will look like the following (Fig. 3.11).

28	Colorado Rockies	760	965	162	63.6
29	Florida Marlins	580	725	162	64.7

Figure 3.11 Excel output involving Colorado Rockies and Florida Marlins of 1993 season.

It's going to be a tough expansion season for both clubs, with the formula projecting 63.6 wins for the Rockies and 64.7 for the Marlins. Based on their runs scored and allowed, both teams are expected to lose nearly 100 games in their inaugural MLB campaigns.

Mini Project 1

This activity will guide you through step-by-step projects.

BASEBALL FOUNDATION (MLB 1979 SEASON)

1. Open the file MLB Standings 1979.xlsx.
2. Use the Pythagorean Theorem formula with an exponent of 1.83 to calculate each team's expected wins.
3. Create a scatter plot comparing actual wins to expected wins.
4. Add a trendline and R^2 value. What does the R^2 tell you about how well the formula explains team performance?

EXPANSION PROJECTIONS

1. Add the projected runs scored and allowed for the 1993 Colorado Rockies and Florida Marlins to the spreadsheet.
2. Use the formula to project each team's expected number of wins.
3. How realistic do these projections seem for new teams?

While baseball is where the Pythagorean Theorem first gained traction, the logic also applies to other sports. Runs, goals, or points scored and allowed are consistent measures of performance. By adjusting the exponent, the formula adapts to a variety of each sport. Let's see how it works in basketball, football, hockey, and soccer.

THE PYTHAGOREAN THEOREM OF SPORTS

Bill James originally developed the Pythagorean Theorem formula to evaluate team performance in baseball. The beauty of the formula is that it can also be applied to other sports. In each case, the core idea remains the same: team winning percentage can be estimated by the number of goals or points scored and allowed. Let's explore how this simple, yet powerful formula can be used across a variety of sports.

Basketball

The exponent used in the Pythagorean Theorem for the National Basketball Association (NBA) is approximately 13.91, which is one of the highest across team sports. The higher the exponent, the smaller the role that randomness or chance tends to play in game outcomes (Butt, 2022). In other words, better teams in the NBA are much more likely to consistently beat weaker teams. For example, in the 2024–25 regular season, the Cleveland Cavaliers finished with a 64–18 record, while the Washington Wizards and Charlotte Hornets ended at 18–64 and 19–63, respectively. As expected, Cleveland defeated both teams in all eight matchups. This outcome aligns with the high predictability suggested by the sport's Pythagorean exponent.

To apply the Pythagorean Theorem to the NBA, download the file "NBA 2024–25 Standings.xlsx" from the book's website. Basketball uses points rather than runs. In the spreadsheet, you will find that PF is points for and PA is points against. In Excel, in cell H2, the formula for expected wins (X-W) is:

$$= (E2\hat{}13.91/(E2\hat{}13.91 + F2\hat{}13.91)) \star G2$$

You can use the steps that were provided in previously in this chapter to create a scatter chart like the following (Fig. 3.12).

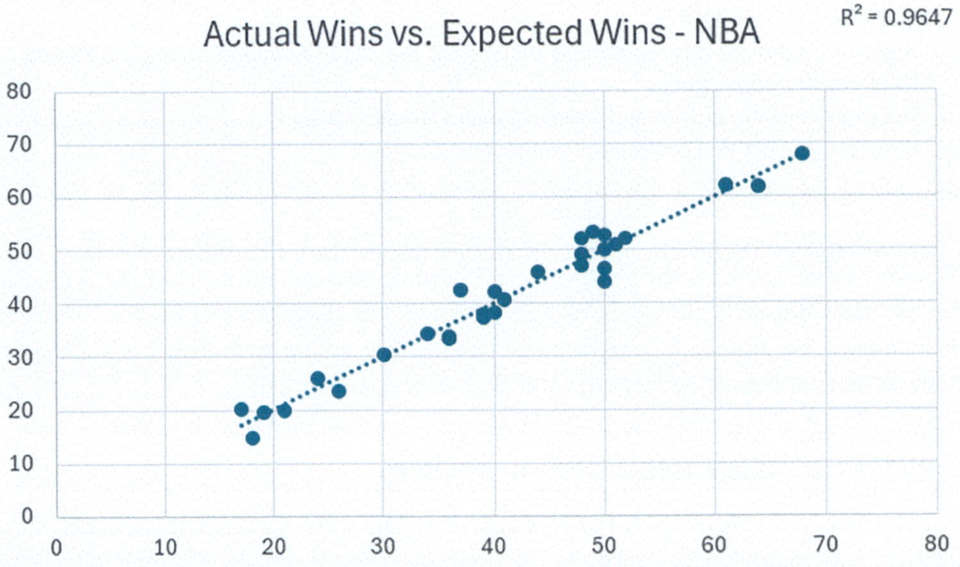

Figure 3.12 Actual wins versus expected wins scatter chart for the NBA 2024–25 season with R-squared coefficient.

The Women's National Basketball Association (WNBA) has an exponent close to 10 (Basketball-Reference, n.d.). This also reflects that there is not much randomness in the WNBA.

To apply the Pythagorean Theorem to the WNBA, download the file "WNBA 2024 Standings.xlsx" from the book's website. In Excel, in cell H2, the formula for expected wins (X-W) is:

= (E2^10/(E2^10 + F2^10)) * G2

Here is a scatter chart from WNBA data (Fig. 3.13).

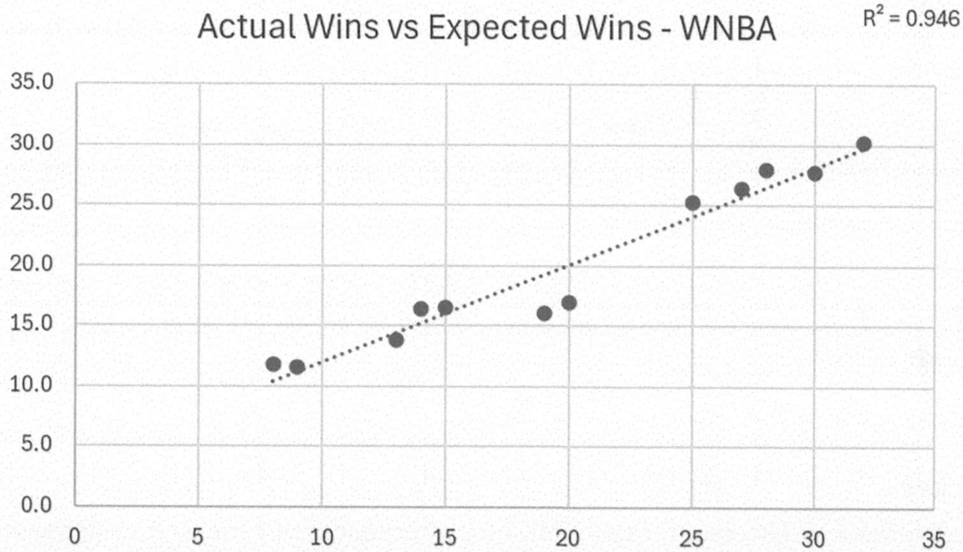

Figure 3.13 Actual wins versus expected wins scatter chart for the WNBA 2024–25 season with R-squared coefficient.

Football

The NFL has an exponent of close to 2.37 (Zivkovic, 2020). Just like the saying "on any given Sunday," as demonstrated by the low exponent used for the NFL Pythagorean Theorem, anything can happen in the NFL. This was no more evident than on November 12, 2023, when five games were decided by a field goal as time expired (Henderson, 2023).

To apply the Pythagorean Theorem to the NFL, download the file "NFL 2024–25 Standings.xlsx" from the book's website. In Excel, in cell H2, the formula for expected wins (X-W) is:

= (E2^2.37/(E2^2.37 + F2^2.37)) * G2

Below is a scatter chart based on the NFL data and the Pythagorean Theorem using an exponent of 2.37 (Fig. 3.14).

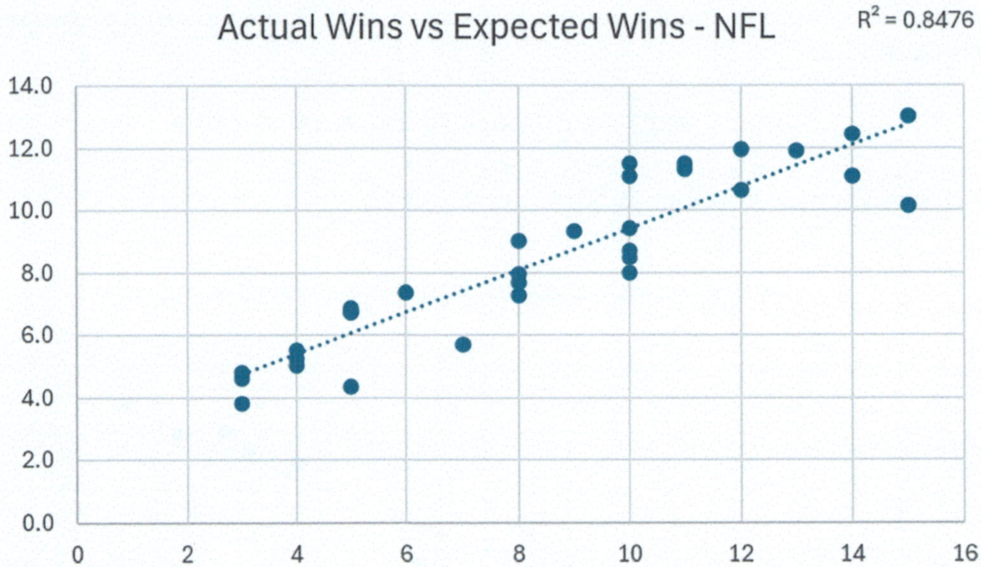

Figure 3.14 Actual wins versus expected wins scatter chart for the NFL 2024–25 season with R-squared coefficient.

Hockey

The Pythagorean Theorem for the NHL uses an exponent of close to 2.15 (Zivkovic, 2020). As one would expect, a sport played on ice has a considerable amount of randomness. To apply the Pythagorean Theorem to the NHL, download the file "NHL 2024–25 Standings.xlsx" from the book's website. In Excel, in cell H2, the formula for expected wins (X-W) is:

= (E2^2.15/(E2^2.15 + F2^2.15)) * G2

Below is a scatter chart based on the NHL data using the Pythagorean Theorem (Fig. 3.15).

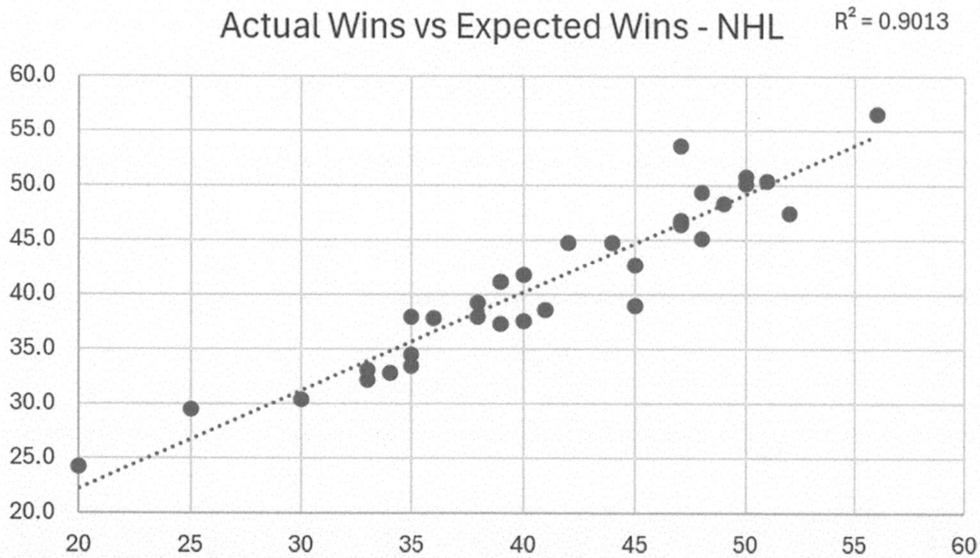

Figure 3.15 Actual wins versus expected wins scatter chart for the NHL 2024–25 season with R-squared coefficient.

Soccer

Due to the closeness and low-scoring nature of games, it's not surprising that the exponent used in the Pythagorean Theorem for the English Premier League would be low. An exponent of just 1.3 for the EPL suggests a high level of randomness in match outcomes (Zivkovic, 2020). To apply the Pythagorean Theorem to the EPL, download the file "EPL 2024–25 Standings.xlsx" from the book's website. In Excel, in cell H2, the formula for expected wins (X-W) is:

$$= (E2\hat{\ }1.3/(E2\hat{\ }1.3 + F2\hat{\ }1.3)) * G2$$

Below is a scatter chart based on the EPL data using the Pythagorean Theorem (Fig. 3.16).

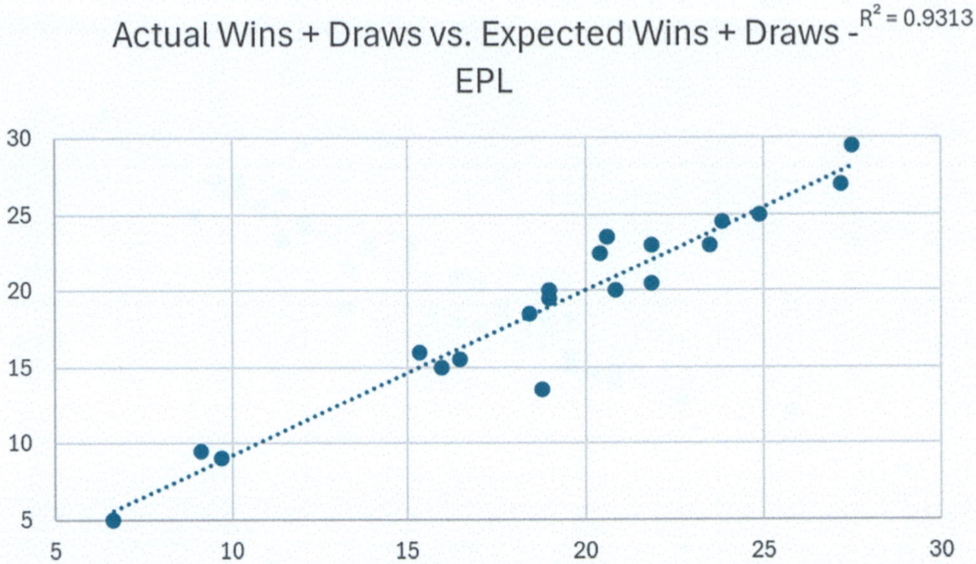

Figure 3.16 Actual wins versus expected wins scatter chart for the EPL 2024–25 season with R-squared coefficient.

It's important to note that, in soccer, draws are common and are typically counted as half a win and half a loss. To account for this, an additional column titled W+D (wins plus draws) was added to column I of the spreadsheet. The scatter chart above compares expected wins plus draws (X-W+D) to actual wins plus draws (W+D).

The exponent of 1.7 used for Major League Soccer (MLS) is low, but not as low as that which was used for the EPL (Hamilton, 2010). To apply the Pythagorean Theorem to the MLS, download the file "MLS 2024–25 Standings.xlsx" from the book's website. In Excel, in cell H2, the formula for expected wins plus draws (X-W+D) is:

= (E2^1.7/(E2^1.7 + F2^1.7)) * G2

Below is a scatter chart based on the MLS data using the Pythagorean Theorem (Fig. 3.17).

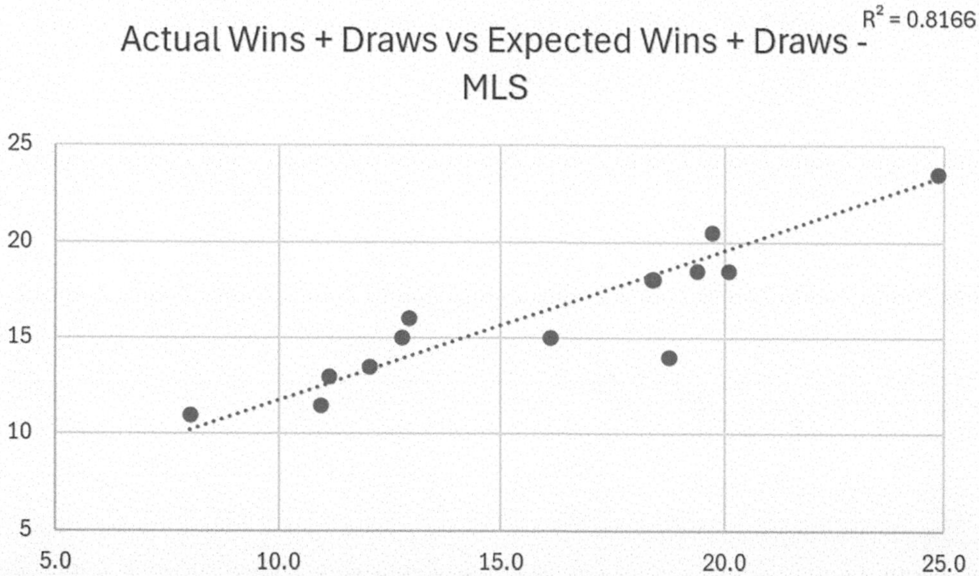

Figure 3.17 Actual wins versus expected wins scatter chart for the MLS 2024–25 season with R-squared coefficient.

One team that is an outlier from the trendline is the Philadelphia Union. Despite having a positive goal differential of +7, Philadelphia had a record of just 9–15–10. In 2024, 12 of the Union's 15 losses were by just one goal. This bad luck cost Philadelphia head coach Jim Curtin his job after 11 seasons with the team (Associated Press, 2024).

To analyze the National Women's Soccer League (NWSL), use an exponent of 1.45 with the Pythagorean Theorem. To apply the formula, download the file "NWSL 2024 Standings.xlsx" from the book's website. In Excel, in cell H2, the formula for expected wins plus draws (X-W+D) is (Fig. 3.18):

= (E2^1.45/(E2^1.45 + F2^1.45)) * G2

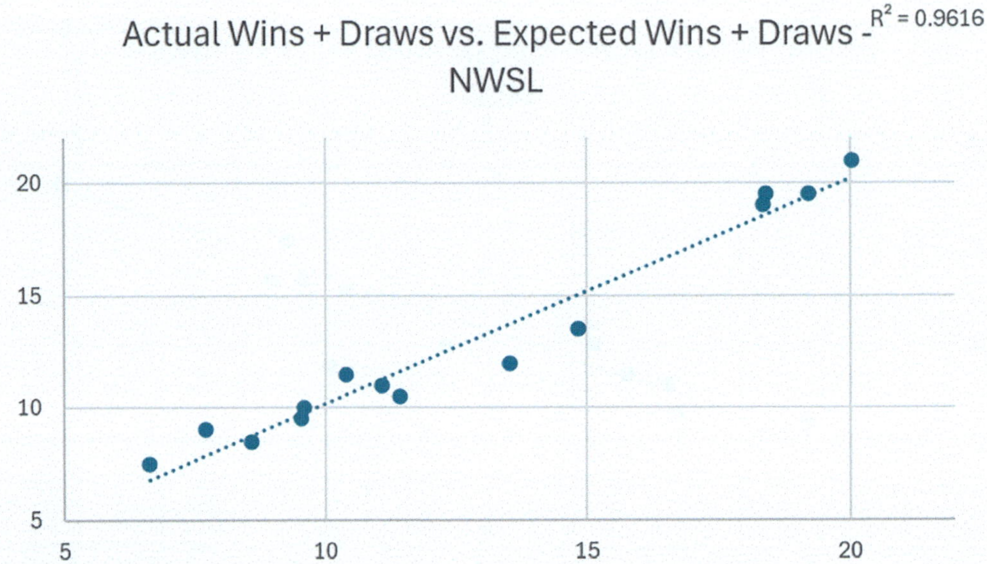

Figure 3.18 Actual wins versus expected wins scatter chart for the NWSL 2024–25 season with R-squared coefficient.

Mini Project 2

1. Download one of the following datasets: NBA 2024–25 Standings.xlsx, NFL 2024–25 Standings.xlsx, or EPL 2024–25 Standings.xlsx.
2. Apply the Pythagorean Theorem with the sport-specific exponent.
3. Create a scatter plot with a trendline and R^2.
4. Choose one team from your dataset where the expected wins differ most from actual wins. What might explain this difference (close games, injuries, or randomness)?

SUMMARY

This chapter shows how to use Bill James' Pythagorean Theorem to predict how many games a team should have won based on how many runs or points were scored and allowed. As a tool for predicting and evaluating team performance, the Pythagorean Theorem of Baseball helped establish James' legacy as a trailblazer in sport analytics. This formula opened the door for coaches, teams, and fans to use data to evaluate how well a team performed, separating skill from luck, and setting the stage for advances in sport analytics. This chapter also demonstrates how Bill James' Pythagorean Theorem can be adapted beyond baseball to estimate team performance across various sports. By using sport-specific exponents, the formula helps predict winning percentages in leagues like the NBA, WNBA, NFL, NHL, EPL, MLS, NWSL, and more.

BEYOND THE BOX SCORE

1. Bill James emphasized breaking down complex questions into simpler, answerable ones. How does using a tool like Excel help you do this in sport analytics? In what ways does it help move from theory to application?
2. The Pythagorean Theorem of Baseball projected team wins with a high R^2 value of 0.9202. What does this tell you about the model's ability to explain team performance? What are the limitations of using R^2 alone when evaluating accuracy?
3. Some teams performed better or worse than predicted, such as the 1979 MLB examples or the Philadelphia Union in soccer. What might explain why teams win more or fewer games than expected? What other types of data would you analyze to investigate these differences?
4. Each sport has a different exponent, reflecting the balance of predictability and randomness. How does having a high exponent (e.g., NBA) versus a low exponent (e.g., soccer) influence how teams, analysts, or even gamblers might view and approach these sports?
5. Reflect on the practical value of the Pythagorean Theorem in professional sport organizations. How might teams use this model differently across sports, and how should analysts balance model projections with real-world factors such as chemistry, injuries, or coaching changes?

AI DISCLOSURE

During the preparation of this work, the author(s) used ChatGPT (OpenAI GPT-5) in order to provide editing and improve the readability of the document. After using this tool/service, the author(s) reviewed and edited the content as needed and take(s) full responsibility for the content of the publication.

REFERENCES

Butt, M. M., 2022. Pythagorean expectation in sports analytics, with examples from different sports. Available from: https://medium.com/data-science/pythagorean-expectation-in-sports-analytics-with-examples-from-different-sports-f5e599530a6c.

Basketball-Reference, n.d. Glossary. Available from: https://www.basketball-reference.com/about/glossary.html (accessed 03.06.25).

Hamilton, H., 2010. Pythagorean 2.0 projections: 2010 MLS regular season. *Soccermetrics*. Available at: https://www.soccermetrics.net/soccer-pythagorean-tables/pythagorean-2010-mls-20100822 (accessed 29.06.26).

Henderson, B., 2023. 5 NFL games end on game winning field goals, most in one day. *ESPN.* Available at: https://www.espn.com/nfl/story/_/id/38886957/record-5-nfl-games-end-game-winning-field-goals (accessed 29.06.2026).

James, B., 2015. Bill James Handbook 2016. United States: ACTA Sports.

Associated Press, 2024. 'MLS' Philadelphia Union fires coach Jim Curtin after 11 years. *ESPN.* Available at: https://www.espn.com/soccer/story/_/id/42250589/mls-philadelphia-union-fires-coach-jim-curtin (accessed 29.06.2026).

Zivkovic, J., 2020. Applying Pythagorean expectation to major sports [Kaggle notebook]. Kaggle. Available at: https://www.kaggle.com/code/jaseziv83/applying-pythagorean-expectation-to-major-sports (accessed 03.06.25).

CHAPTER 4

Correlation

Key Takeaways

- Correlation measures the strength and direction of the relationship between two variables.
- The correlation coefficient (–1 to 1) helps identify whether skills are consistent, unrelated, or move in opposite directions.
- Some performance metrics are more repeatable than others, making them stronger predictors of future success across seasons.

Ultimately, all science is correlation.

– Peter Watts (Watts, 2014)

In sport analytics, we often want to know if two variables move together. Correlation helps answer important questions like:

- Which player skills are consistent year over year?
- Are some stats better predictors of future success than others?
- Can we trust that a free agent will keep performing at their current level?

Correlation describes the strength and direction of the relationship between two variables. For example, a positive correlation might exist between batting average and on-base percentage. In this relationship, when one goes up (batting average), the other tends to go up too (on-base percentage). A negative correlation could be seen between earned run average (ERA) and win percentage. Here, when ERA rises, win percentage usually drops. Sometimes, two variables show no meaningful relationship. For example, a player's batting average and their fielding percentage lack correlation since a good hitter isn't necessarily a good fielder, and vice versa.

We measure correlation using a correlation coefficient, which ranges from −1 to 1. A value close to 1 means a strong positive relationship, −1 indicates a strong negative relationship, and 0 means no correlation. To use this method properly, the data should:

- Be continuous (like time, distance, or percentage).
- Follow a normal distribution. This means that the data forms a bell-shaped curve, where most values cluster around the average with fewer values at the extremes.
- Reflect a linear relationship (i.e., follow a roughly straight-line pattern).
- Be free from extreme outliers that could distort the results.

An Introduction to Applied Sport Analytics **https://doi.org/10.1016/B978-0-443-49068-2.00004-5**
© 2027 Elsevier Inc. All rights are reserved, including those for text and data mining, AI training, and similar technologies.

Excel's Analysis ToolPak makes it easy to perform correlation analysis. This tool allows you to quickly test relationships between two sport variables and decide which ones are worth further analysis.

INSTALLING THE ANALYSIS TOOLPAK

To add the Analysis ToolPak add-in to Excel:

1. Click the File tab, click Options, and then click the Add-Ins category.
2. In the Manage box, select Excel Add-ins and then click Go (Fig. 4.1).

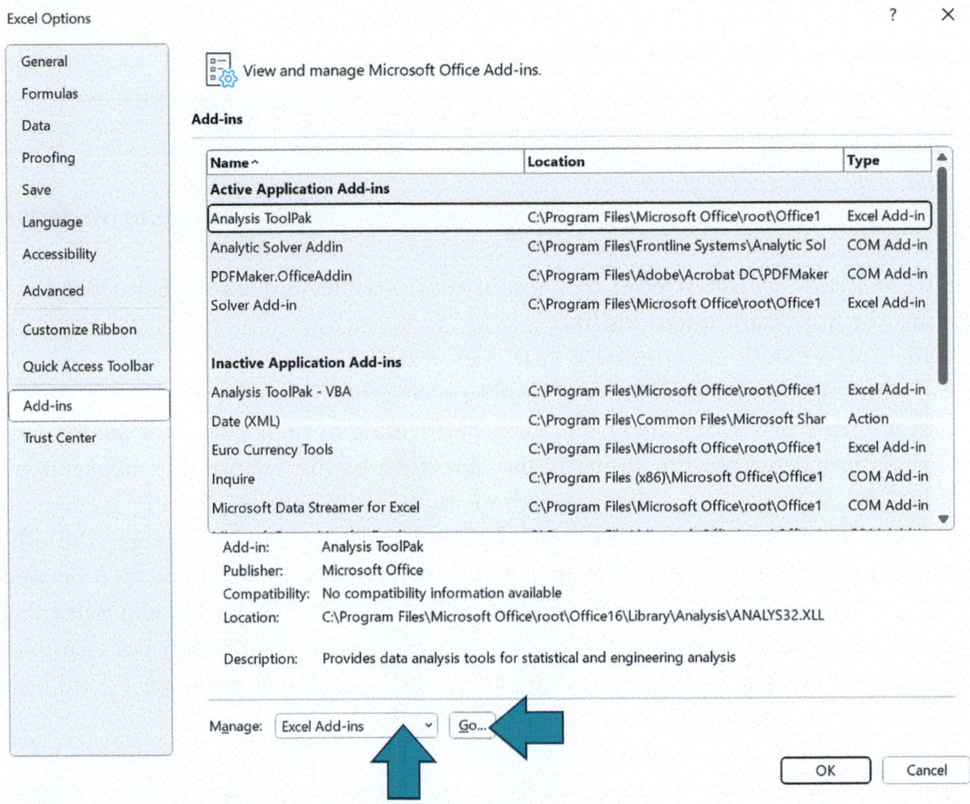

Figure 4.1 ***Add-ins with Analysis ToolPak.*** Add-ins with Analysis ToolPak using Excel on PC.

If you're using Excel for Mac, in the File menu go to Tools > Excel Add-ins.

3. In the Add-ins box, check the Analysis ToolPak check box, and then click OK (Fig. 4.2).

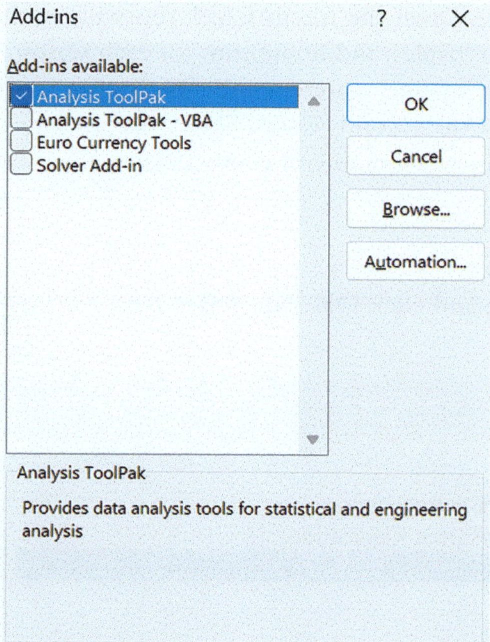

Figure 4.2 The Add-ins box in Microsoft Excel.

Note: If you are prompted with a message that the Analysis ToolPak is not currently installed on your computer, click Yes to install it.

4. Data Analysis is now available on the Data tab (Fig. 4.3).

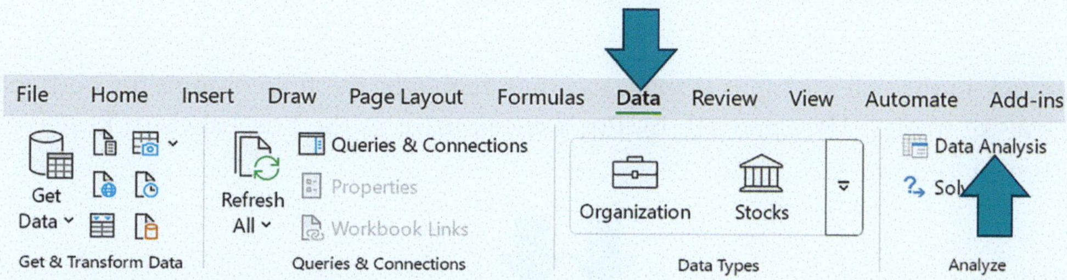

Figure 4.3 ***Data Analysis.*** Data Analysis on the Data tab.

Note: If these instructions don't apply to your computer or if you're having issues, search the Internet for "Load the Analysis ToolPak in Excel." Microsoft offers additional instructions for installing Analysis ToolPak.

RUNNING A CORRELATION

Let's walk through a hands-on example using Excel and player data from two consecutive MLB seasons. Download the file "MLB Correlation Data.xlsx" (Lahman, n.d.)

from the book's website. Open the file in Excel. You will see that the dataset includes stats like singles, doubles, triples, and home runs for each season. Columns labeled with an "_a" and colored in orange represent the first season; columns labeled "_b" and colored in green are for the second season. The dataset includes over 400 players who had at least 500 plate appearances in consecutive years.

In Excel:

1 Click the Data tab.
2 Select Data Analysis.
3 Choose Correlation and click OK (Fig. 4.4).

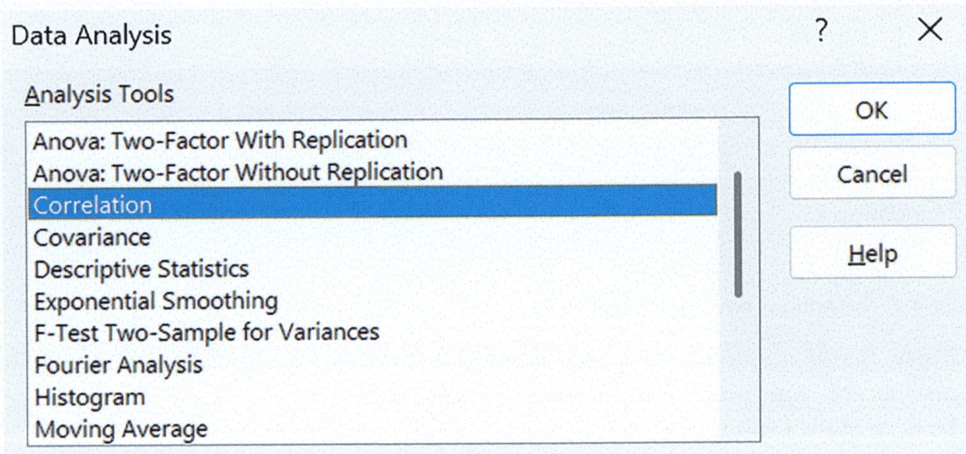

Figure 4.4 ***Correlation.*** Highlighting of correlation function.

4 Click in the cell for Input Range, and then highlight the full columns for both seasons of singles (1B_a and 1B_b). Put a check mark in the box Labels in First Row (Fig. 4.5).

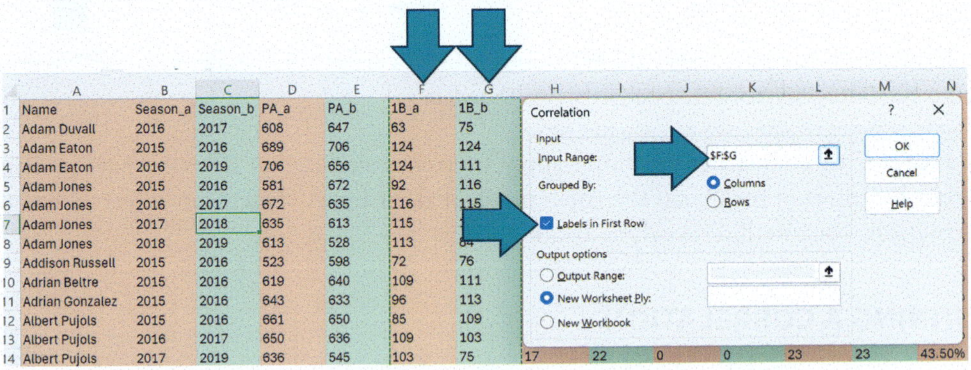

Figure 4.5 ***Input Range.*** Arrows demonstrating what is being written into the Input Range.

5 Click OK to generate a correlation matrix.

INTERPRETING THE RESULTS

For correlation, it is generally accepted that the following scale can be used to estimate the effect size (Cohen, 1988):

- ±0.5 = large
- ±0.3 = medium
- ±0.1 = small

Displayed on a scatter chart, these results might look like the following (Fig. 4.6).

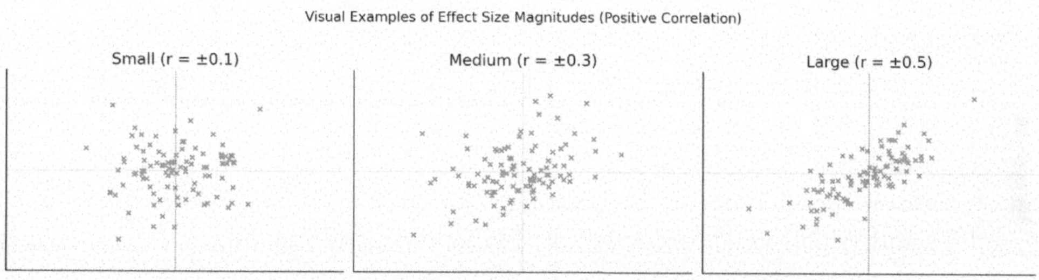

Figure 4.6 ***Scatter Charts.*** Three scatter charts showing different correlations.

The results from our correlation matrix show a correlation of 0.627944 between singles in one season (1B_a) and singles in the following season (1B_b). This represents a large effect size (Fig. 4.7).

	A	B	C
1		*1B_a*	*1B_b*
2	1B_a	1	
3	1B_b	0.627944	1

Figure 4.7 ***Correlation Matrix.*** Correlation matrix showing large effect size.

This result suggests that the number of singles a player hits in one season is strongly associated with the number of singles he is likely to hit in the next season. In other words, hitting singles tends to be a consistent skill across seasons for individual players. If you're evaluating player performance, correlation can help forecast future performance.

Further analysis revealed the following for how hitting statistics correlate from season to season for batters who play a lot in both seasons.

Correlations

Stat pair	Correlation	Interpretation
Singles (A vs. B)	+0.63	Strong positive correlation
Doubles (A vs. B)	+0.42	Medium positive correlation
Triples (A vs. B)	+0.48	Medium positive correlation
Home runs (A vs. B)	+0.62	Strong positive correlation
Singles (A) versus home runs (B)	−0.35	Moderate negative correlation

Key insights from the correlation analysis above include:

- Singles and home runs are largely repeatable skills. Players with over 500 at-bats in one season are likely to perform similarly in these categories in the next season.
- Doubles and triples are less consistent. These stats may be influenced by ballpark factors and randomness.
- The negative correlation between singles and home runs suggests these are different player profiles (e.g., contact hitters vs. power hitters). In other words, don't expect a singles hitter like Luis Arraez to hit home runs in the next season like Kyle Schwarber, and vice versa.

Using the spreadsheet, apply the same method to different metrics, including:

- Ground ball rate
- Line drive rate
- Fly ball rate
- Home runs per fly ball (HR/FB)

What are the takeaways from your analysis?

RUNNING A CORRELATION II—NWSL

To gain additional practice performing correlation analysis, download the file "NWSL Correlation Data.xlsx" from the book's website. This file contains player data for players in the National Women's Soccer League (NWSL). In order to place the statistics on a level playing field, most have been converted into a ratio of per minute. Categories that we will use in our analysis include:

- Goals per minute (G/Min)
- Assists per minute (A/Min)
- Total scoring attempts per minute (S/Min)
- On-target scoring attempts per minute (SOT/Min)
- Accurate pass percentage (Pass%)

- Tackles won per minute (Tackles/Min)
- Fouls committed per minute (FC/Min)

As in our previous file, columns labeled with an "_a" and colored in orange represent the first season (2023); columns labeled "_b" and colored in green are for the second season (2024). The dataset includes 28 players who played at least 1800 minutes in consecutive seasons.

Following the steps we used in the previous example, access the Correlation window. For Input Range, select the data between cells G1 and T29. Then click [OK] (Fig. 4.8).

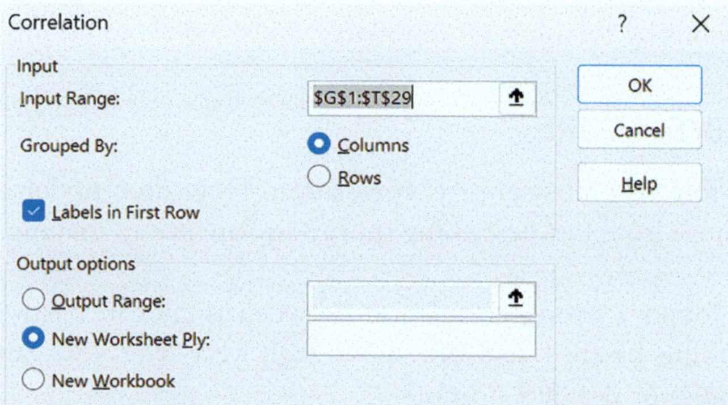

Figure 4.8 ***New Input Range of NWSL data.*** New input range selected from NWSL dataset.

The results below have had conditional formatting added to them. You can see that there is a large correlation effect (greater than 0.5) from one season to the next for total scoring attempts (S/Min), on-target scoring attempts (SOT/Min), tackles won (Tackles), and fouls committed (FC). These results suggest that in the NWSL, offensive metrics like total and on-target scoring attempts tend to be consistent year to year for regular starters, while creative or situational stats (like assists) fluctuate more (Fig. 4.9).

	A	B	C	D	E	F	G	H	I	J	K	L	M	N	O
1		G/Min_a	G/Min_b	A/Min_a	A/Min_b	S/Min_a	S/Min_b	SOT/Min_a	SOT/Min_b	Pass%_a	Pass%_b	ckles/Min_	ckles/Min_	FC/Min_a	FC/Min_b
2	G/Min_a	1													
3	G/Min_b	0.399315	1												
4	A/Min_a	-0.12624	0.209477	1											
5	A/Min_b	0.518832	0.691274	0.284529	1										
6	S/Min_a	0.865654	0.6591	0.092725	0.627437	1									
7	S/Min_b	0.698648	0.816709	0.09484	0.666425	0.91397	1								
8	SOT/Min_a	0.873399	0.593161	0.01449	0.598074	0.943845	0.869335	1							
9	SOT/Min_b	0.621772	0.859743	0.165437	0.717126	0.859835	0.951568	0.840527	1						
10	Pass%_a	0.155947	0.075553	-0.03972	0.230892	0.015514	0.001035	0.168969	0.13203	1					
11	Pass%_b	-0.21041	-0.15539	0.210783	-0.08922	-0.21322	-0.28203	-0.25003	-0.21908	-0.00191	1				
12	Tackles/M	0.382539	0.167823	0.485145	0.337678	0.433706	0.359884	0.332597	0.25119	-0.17336	0.033155	1			
13	Tackles/M	0.37101	0.358109	0.433302	0.437727	0.472701	0.470308	0.391294	0.411396	0.039655	-0.03141	0.836272	1		
14	FC/Min_a	0.228361	0.377527	0.416886	0.374782	0.429752	0.441403	0.29807	0.359745	-0.27589	0.222393	0.786439	0.68742	1	
15	FC/Min_b	0.29398	0.55551	0.543561	0.566679	0.514379	0.585018	0.420442	0.556455	-0.01462	0.085604	0.739255	0.808109	0.798616	1

Figure 4.9 ***Correlation Matrix.*** Correlation matrix of various statistics of the NWSL.

How could a general manager in charge of player acquisition use these findings when evaluating the free agent market?

SUMMARY

This chapter introduces correlation as a tool in sport analytics for measuring the relationship between two variables. Using Excel's Analysis ToolPak with MLB and NWSL data, it shows that some skills are consistent across seasons, while other stats are less reliable. Understanding correlation strength (small, medium, or large) helps analysts identify which player metrics are most predictive of future performance. This enables teams to make better decisions by focusing on repeatable skills and avoiding misleading data.

BEYOND THE BOX SCORE

1. Peter Watts wrote that all science is correlation. How does understanding correlation help us move from just describing sport performance to making evidence-based decisions?
2. This chapter found a strong correlation between singles in back-to-back seasons. Why might some baseball statistics show high year-over-year consistency while others, like triples or doubles, do not?
3. There was a negative correlation between singles and home runs. What does this tell you about player types and how teams might use this information when constructing a lineup or making roster decisions?
4. In the NWSL analysis, which metrics showed the greatest year-to-year consistency? Why might these skills translate more reliably across seasons than others?

Mini Project

1. Run a correlation analysis on either the MLB or NWSL dataset introduced in this chapter.
2. Interpret your findings. Think about why some statistics might be more consistent across some seasons than others. Consider player roles, randomness, or game context.
3. Write a letter of 250–300 words to a general manager in the league. Based on your analysis, recommend which performance metrics should be prioritized in player evaluation, contract decisions, or scouting. Explain how understanding repeatable skills can provide a competitive advantage.

5. How could an NWSL general manager use correlation results to inform player signings or contract renewals? For example, should a team prioritize players with consistently high scoring attempts or take a chance on players with fluctuating but high-upside metrics like assists?

AI DISCLOSURE

During the preparation of this work, the author(s) used ChatGPT (OpenAI GPT-5) in order to provide editing and improve the readability of the document. After using this tool/service, the author(s) reviewed and edited the content as needed and take(s) full responsibility for the content of the publication.

REFERENCES

Cohen, J., 1988. Statistical power analysis for the behavioral sciences. 2.
Lahman, S. Lahman Baseball Database. Available from: https://www.seanlahman.com.
Watts, P., 2014. Echopraxia. 2.

CHAPTER 5

Simple Linear Regression

Key Takeaways

- Simple linear regression helps explain how one variable affects another variable.
- The regression equation's key statistics, *P*-value and R^2, indicate how strong and meaningful the relationship is.
- Understanding regression helps sport managers predict outcomes and make data-driven decisions.

You are what your record says you are.

– Bill Parcells (Parcells and Demasio, 2014)

Linear regression analysis is a statistical analysis tool where one or more independent variables are evaluated for a correlation to a dependent variable. The output of linear regression provides the correlation between variables, much like was described in Chapter 4. Linear regression is a widely used technique in data analysis and a common algorithm in machine learning. Simple linear regression gets its name because it models the relationship between just one independent variable and one dependent variable. When additional independent variables are included, it is referred to as multiple linear regression (see Chapter 6).

Regression has been employed in numerous studies to examine the factors influencing attendance at events. Various sports have been studied for attendance factors using regression, including baseball, basketball, football, hockey, mountain biking, trail running, and soccer. Regression has also been used to analyze attendance variables associated with concerts, sport tourism, and the impact of the internet and social media on event attendance.

The goal of this chapter is to provide an understanding of the application of regression to sport and the business of sport. The chapter outlines the use of regression analysis with attendance data, presenting a step-by-step guide in Excel to help you grasp the relationship between variables and the significance of data analysis.

An Introduction to Applied Sport Analytics **https://doi.org/10.1016/B978-0-443-49068-2.00005-7**
© 2027 Elsevier Inc. All rights are reserved, including those for text and data mining, AI training, and similar technologies.

To better understand simple linear regression, let's look at the relationship between event attendance and winning in college football. Attendance at sporting events is crucial to the financial health of major college athletic departments, as these departments generate much of their revenue through ticket sales, sponsorships, and television rights. Predicting event attendance is valuable for several reasons beyond revenue generation, including efficient staffing, facility management, and marketing efforts.

For this example of simple linear regression, research data on attendance and overall winning percentages for major college football teams were collected from https://stats.ncaa.org (National Collegiate Athletic Association, n.d.). Attendance was measured as a percentage of each school's stadium capacity for home games.

RUNNING A REGRESSION ANALYSIS

Creating a regression analysis in Excel requires the Data Analysis add-in. To add the Data Analysis add-in to Excel, see Chapter 4. After adding the add-in to Excel:

1. Download the file "College Football Regression.xlsx" from the book's website.
2. Open the file in Excel. In addition to school names (Institution), the file also contains data for the average stadium capacity for home games (% capacity) and for the team's winning percentage that season (Win%) (Fig. 5.1).

	A	B	C	D
1	Institution	% Capacity	Win%	Predicted Capacity
2	Pittsburgh	63.41	0.615	
3	UCLA	54.39	0.333	
4	Illinois	60.31	0.462	
5	California	67.93	0.615	
6	Rutgers	57.35	0.167	
7	Florida St.	67.9	0.462	
8	Arizona St.	76.53	0.615	
9	Southern California	76.59	0.615	
10	Virginia	77.83	0.643	
11	Indiana	77.92	0.615	
12	Kansas	67.41	0.250	
13	Louisville	82.09	0.615	
14	Maryland	70.02	0.250	

Figure 5.1 ***College Football Data.*** Institution, average stadium capacity, team win percentage, and predicted capacity.

3. From the Data tab, select Data Analysis.
4. For Analysis Tools, select Regression and click OK (Fig. 5.2).

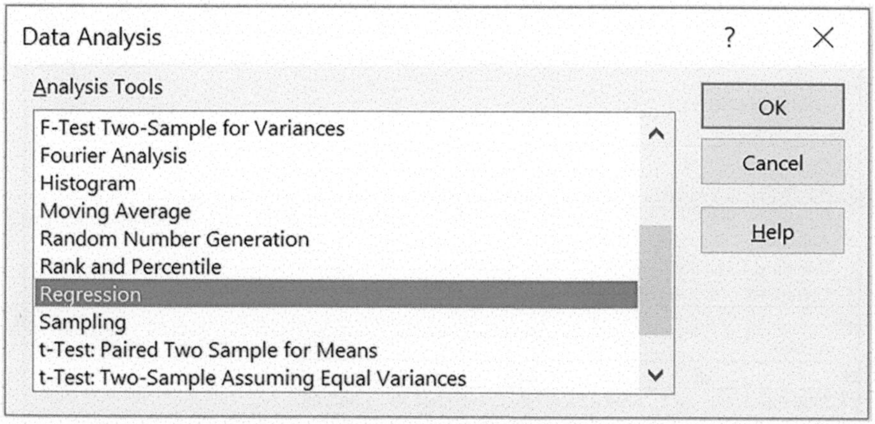

Figure 5.2 ***Regression.*** Regression selected from Analysis Tools.

5. For Input Y Range, drag the cursor over all of the Percent Capacity column B. This is our dependent variable or the outcome we are trying to predict based on other factors.
6. For Input X Range, drag the cursor over all of the Win% in column C. This is our independent variable or the factor we believe may help explain or predict changes in the dependent variable.
   Note: If you also dragged the cursor over the column headings, click the box next to Labels (Fig. 5.3).

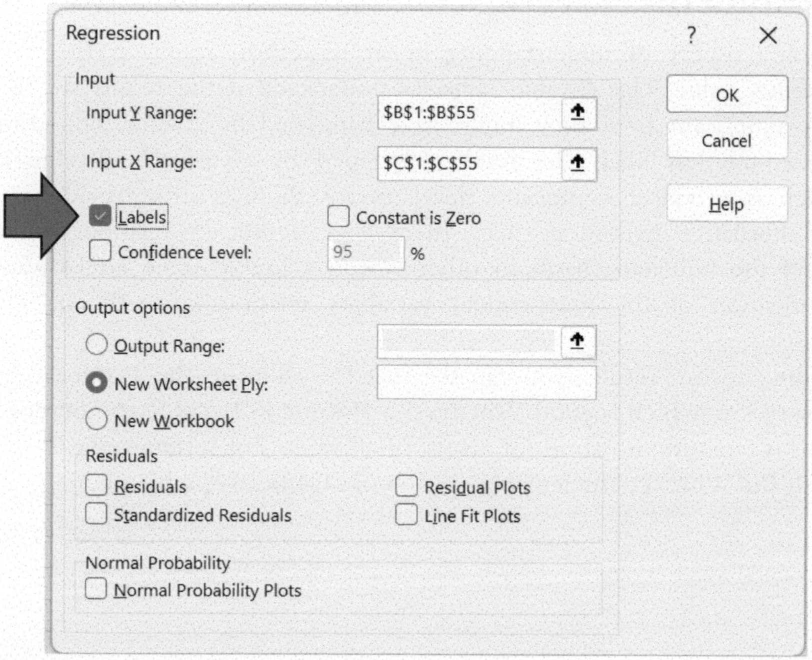

Figure 5.3 ***Regression Labels.*** Regression window in Microsoft Excel.

7. Click the OK button.
8. Excel will return the following regression results (Fig. 5.4).

	A	B	C	D	E	F	G	H	I
1	SUMMARY OUTPUT								
2									
3	*Regression Statistics*								
4	Multiple R	0.552513							
5	R Square	0.305271							
6	Adjusted F	0.29191							
7	Standard I	11.06855							
8	Observatio	54							
9									
10	ANOVA								
11		*df*	*SS*	*MS*	*F*	*gnificance F*			
12	Regressior	1	2799.329	2799.329	22.84928	1.48E-05			
13	Residual	52	6370.666	122.5128					
14	Total	53	9169.995						
15									
16		*Coefficients*	*andard Err*	*t Stat*	*P-value*	*Lower 95%*	*Upper 95%*	*ower 95.0%*	*pper 95.0%*
17	Intercept	67.76848	4.334848	15.63342	2.69E-21	59.06997	76.46699	59.06997	76.46699
18	Win%	34.12738	7.139482	4.780092	1.48E-05	19.80096	48.4538	19.80096	48.4538

Figure 5.4 ***Regression Results.*** Regression results with statistics and coefficients.

INTERPRETING THE RESULTS

An important aspect of understanding linear regression results is the *P*-value, or the probability value. The *P*-value helps us understand if the results we receive are statistically significant or if they might have happened by chance. Thus the smaller the *P*-value, the less likely the results happened by chance. A *P*-value less than 0.05, which is a typical significance level, means there is strong evidence to support the alternative hypothesis (e.g., there is a significant relationship), allowing us to reject the null hypothesis. In other words, a low *P*-value could lead to further investigation of the independent variable's relationship with the dependent variable.

According to our results, you can see that the value in the *P*-value column for Win% is very low, which is good. Whenever a *P*-value in Excel has a notation with an E in it, this is scientific notation for a very small value. For instance, the *P*-value associated with the Win% coefficient has a value of 1.48×10^{-5}, which equals 0.0000148 (Fig. 5.5).

	A	B	C	D	E	F	G	H	I
1	SUMMARY OUTPUT								
2									
3	*Regression Statistics*								
4	Multiple R	0.552513							
5	R Square	0.305271							
6	Adjusted F	0.29191							
7	Standard I	11.06855							
8	Observati	54							
9									
10	ANOVA								
11		*df*	*SS*	*MS*	*F*	*gnificance F*			
12	Regressior	1	2799.329	2799.329	22.84928	1.48E-05			
13	Residual	52	6370.666	122.5128					
14	Total	53	9169.995						
15									
16		*Coefficients*	*andard Err*	*t Stat*	*P-value*	*Lower 95%*	*Upper 95%*	*ower 95.0%*	*pper 95.0%*
17	Intercept	67.76848	4.334848	15.63342	2.69E-21	59.06997	76.46699	59.06997	76.46699
18	Win%	34.12738	7.139482	4.780092	1.48E-05	[illegible]	48.4538	19.80096	48.4538

Figure 5.5 ***P-value.*** *P*-value of slope coefficient.

The results also provide coefficients for Intercept and Win%. Intercept represents the constant in our formula, meaning it remains fixed regardless of changes in Win%. We will apply these coefficients to estimate attendance (Fig. 5.6).

	A	B	C	D	E	F	G	H	I
1	SUMMARY OUTPUT								
2									
3	*Regression Statistics*								
4	Multiple R	0.552513							
5	R Square	0.305271							
6	Adjusted F	0.29191							
7	Standard I	11.06855							
8	Observati	54							
9									
10	ANOVA								
11		*df*	*SS*	*MS*	*F*	*gnificance F*			
12	Regressior	1	2799.329	2799.329	22.84928	1.48E-05			
13	Residual	52	6370.666	122.5128					
14	Total	53	9169.995						
15									
16		*Coefficients*	*andard Err*	*t Stat*	*P-value*	*Lower 95%*	*Upper 95%*	*ower 95.0%*	*pper 95.0%*
17	Intercept	67.76848	[illegible]	15.63342	2.69E-21	59.06997	76.46699	59.06997	76.46699
18	Win%	34.12738	[illegible]	4.780092	1.48E-05	19.80096	48.4538	19.80096	48.4538

Figure 5.6 ***Coefficients of Regression Analysis.*** Arrows identifying the coefficients of the linear regression model.

APPLYING THE REGRESSION RESULTS

To calculate the Predicted Capacity column using our regression results:

1. Click on the sheet in Excel.
   Under the column header Predicted Capacity, in cell D2, type the formula: =67.76848+(34.12738×C2) and press Enter. These are the values taken from the regression results under the column labeled Coefficients. If you click in cell D2, your spreadsheet should contain the following formula. The formula is also displayed in the formula bar at the top of the spreadsheet (Fig. 5.7).

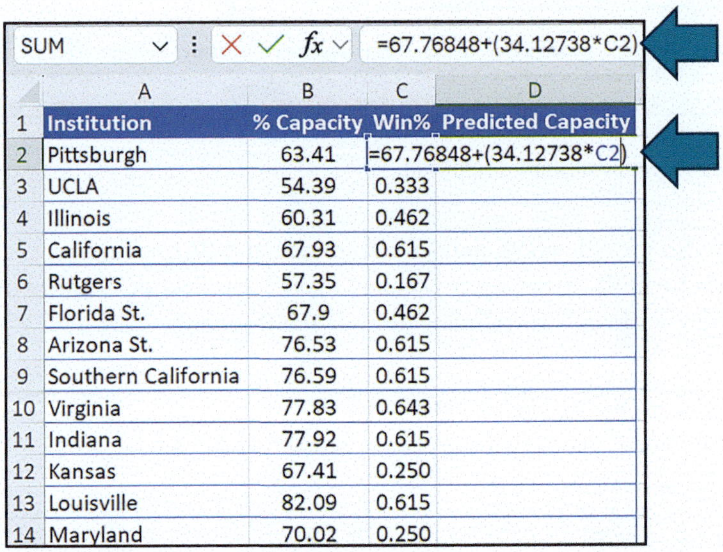

SUM =67.76848+(34.12738*C2)

	A	B	C	D
1	Institution	% Capacity	Win%	Predicted Capacity
2	Pittsburgh	63.41	=67.76848+(34.12738*C2)	
3	UCLA	54.39	0.333	
4	Illinois	60.31	0.462	
5	California	67.93	0.615	
6	Rutgers	57.35	0.167	
7	Florida St.	67.9	0.462	
8	Arizona St.	76.53	0.615	
9	Southern California	76.59	0.615	
10	Virginia	77.83	0.643	
11	Indiana	77.92	0.615	
12	Kansas	67.41	0.250	
13	Louisville	82.09	0.615	
14	Maryland	70.02	0.250	

Figure 5.7 ***Predicted Capacity.*** Application of regression results to calculate predicted capacity.

2. Left double-click the box in the lower right corner of cell D2 to populate the entire column, or left click and drag down to cell D55. The projected attendance values will populate in column D (Fig. 5.8).

	A	B	C	D
1	Institution	% Capacity	Win%	Predicted Capacity
2	Pittsburgh	63.41	0.615	88.8
3	UCLA	54.39	0.333	79.1
4	Illinois	60.31	0.462	83.5
5	California	67.93	0.615	88.8
6	Rutgers	57.35	0.167	73.5
7	Florida St.	67.9	0.462	83.5
8	Arizona St.	76.53	0.615	88.8
9	Southern California	76.59	0.615	88.8
10	Virginia	77.83	0.643	89.7
11	Indiana	77.92	0.615	88.8
12	Kansas	67.41	0.250	76.3
13	Louisville	82.09	0.615	88.8
14	Maryland	70.02	0.250	76.3
15	Minnesota	90.92	0.846	96.6
16	Arizona	73.69	0.333	79.1
17	Stanford	74.04	0.333	79.1
18	Ole Miss	75.32	0.333	79.1

Figure 5.8 ***Filled in Predicted Capacity.*** Filled in predicted capacity.

To create a scatter chart showing the relationship between the actual percentage of capacity and the predicted percentage of capacity, in Excel:

1. Highlight the columns % Capacity and Predicted Capacity. To highlight both columns, hold the CTRL key with a left click and drag to highlight these two columns.
2. From the Insert tab, click the drop-down for Scatter Charts and choose Scatter (Fig. 5.9).

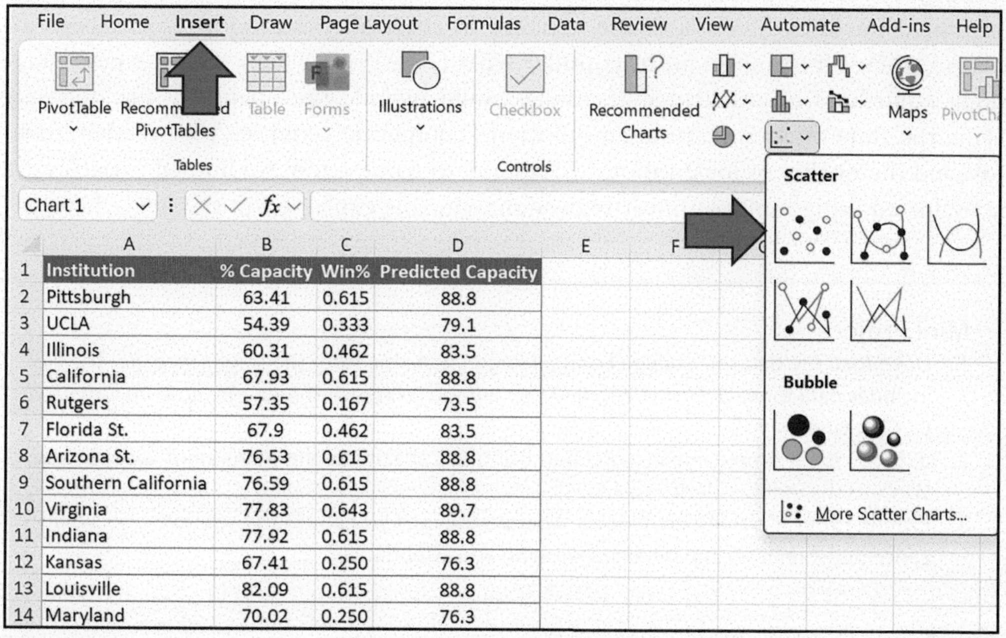

	A	B	C	D
1	Institution	% Capacity	Win%	Predicted Capacity
2	Pittsburgh	63.41	0.615	88.8
3	UCLA	54.39	0.333	79.1
4	Illinois	60.31	0.462	83.5
5	California	67.93	0.615	88.8
6	Rutgers	57.35	0.167	73.5
7	Florida St.	67.9	0.462	83.5
8	Arizona St.	76.53	0.615	88.8
9	Southern California	76.59	0.615	88.8
10	Virginia	77.83	0.643	89.7
11	Indiana	77.92	0.615	88.8
12	Kansas	67.41	0.250	76.3
13	Louisville	82.09	0.615	88.8
14	Maryland	70.02	0.250	76.3

Figure 5.9 ***Insert Scatter Chart.*** Illustration to show how to insert a scatter chart.

3. To add a trendline, right-click any of the data points and choose Add Trendline.
4. Use the check boxes to add the equation or the R-squared value to the chart (Fig. 5.10).

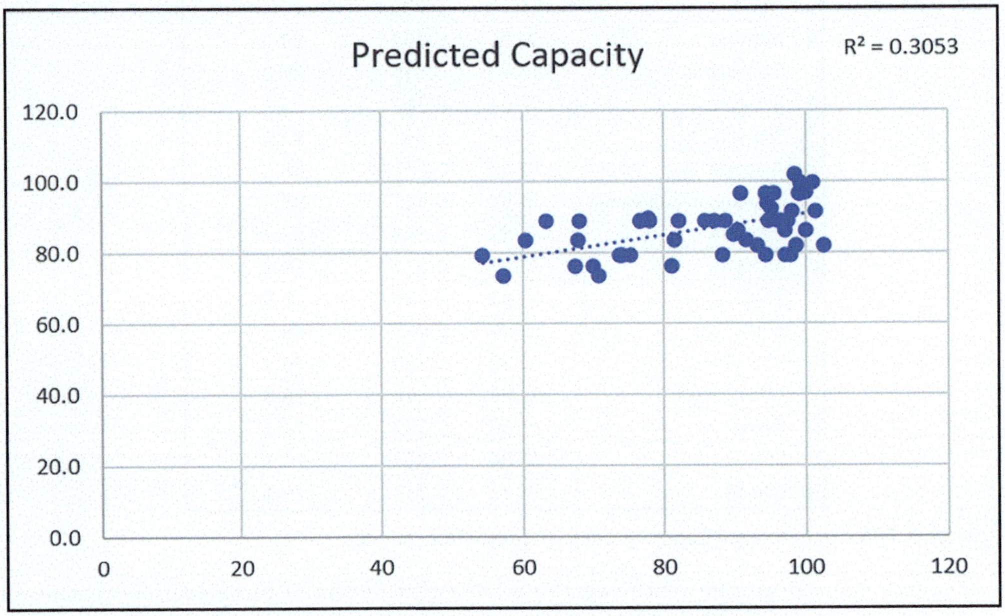

Figure 5.10 ***Scatter Chart.*** Scatter chart of winning percentage and attendance.

Our research suggests there is a relationship between winning percentage and attendance. The analysis revealed an R^2 of 0.3043, meaning that winning percentage accounted for just over 30% of the variance in attendance according to the percentage of capacity. It is important to consider what factors other than winning percentage might contribute to attendance. Examples could include the quality of the opposing teams, the time or date of games, promotions, competing activities, the weather, tradition, and the quality or location of the stadium, to name a few. Additional variables can be evaluated using multiple linear regression, which is explained in the next chapter.

Mini Project

1. Download the dataset "College Football Regression.xlsx" from the book's website. The file includes data for each team's Winning Percentage (Win%) and Average Stadium Capacity (% Capacity).
2. In Excel, run a simple linear regression using % Capacity as the dependent variable and Win% as the independent variable.
3. Examine your R^2 value and *P*-value. What do they tell you about the strength and significance of the relationship between winning and attendance?

4. Create a scatter plot showing the actual and predicted attendance values and include the trendline.
5. Write a short (200–300 word) reflection describing what your results suggest about how winning affects fan attendance. Discuss at least one other factor you believe might also influence attendance and explain why it could be important for sport managers to consider.

SUMMARY

This chapter introduced simple linear regression, a method used to explore the relationship between one independent and one dependent variable. Regression is important because it helps us understand and measure how changes in one variable (like team performance) can help predict changes in another (like attendance). In sport analytics, this is especially useful for making informed decisions about marketing, staffing, ticket pricing, facility planning, and player and team performance. It is a key tool for turning data into actionable insights.

BEYOND THE BOX SCORE

1. Sport organizations often assume that winning leads to higher attendance. Based on your analysis, how would you challenge or support that assumption when advising a college athletic department or professional team?
2. The analysis found a statistically significant relationship between winning percentage and attendance, but the R^2 value was only 0.3043. What does this tell you about the relationship between the two variables, and what other factors might better explain attendance?
3. A low *P*-value indicates strong evidence of a relationship between winning and attendance. Why is statistical significance important, and how should decision-makers in sport organizations interpret it when planning marketing or staffing?
4. Regression produces coefficients that allow us to predict outcomes. How might this ability to forecast help athletic departments or professional franchises plan more efficiently and effectively?
5. After working through this Excel activity, what insights did you gain about using regression in sport analytics? How might you apply this tool in other areas of sport business or performance analysis?

LOOKING AHEAD

In the next chapter, you will build on simple linear regression by learning how to use multiple linear regression. You'll see how this method helps explain outcomes influenced by several predictor variables, providing a deeper understanding of relationships within data.

AI DISCLOSURE

During the preparation of this work, the author(s) used ChatGPT (OpenAI GPT-5) in order to provide editing and improve the readability of the document. After using this tool/service, the author(s) reviewed and edited the content as needed and take(s) full responsibility for the content of the publication.

REFERENCES

National Collegiate Athletic Association, n.d. NCAA statistics. https://stats.ncaa.org (accessed 29.06.26).
Parcells, B. and DeMasio, N., 2014. Parcells: A Football Life. United States: Crown.

CHAPTER 6

Multiple Linear Regression

Key Takeaways

- Multiple linear regression looks at how two or more factors together influence an outcome, helping explain complex relationships in sport and business data.
- Coefficients show how much each variable affects the result, while *P*-values tell us if those effects are statistically meaningful.
- Choosing the correct variables is important to ensure accurate results.

In God we trust; all others must bring data.

– Author unknown (widely attributed to W. Edwards Deming; Lohr, 2023)

Multiple linear regression is a statistical method used to understand the relationship between one dependent variable and two or more independent variables. Think of it as a way to see how multiple factors together influence an outcome. For example, a sport organization might use multiple linear regression to figure out how advertising, pricing, and social media impact their ticket sales. In baseball, multiple linear regression helps us find out how different independent variables (like singles, doubles, triples, and home runs) together affect another dependent variable (like total runs scored). We can use baseball statistics with multiple linear regression to create our own formula for analyzing teams or players.

INTERPRETING THE RESULTS

When you perform a multiple linear regression analysis, you get a set of numbers called coefficients that are associated with independent variables. These tell you how much each independent variable (singles, doubles, triples, and home runs) affects the dependent variable (like runs scored). If the coefficient is significant and positive, it suggests there's a direct relationship between more singles, doubles, triples, and home runs, which will lead to more runs scored by a team on average. If it's significant and negative, there's an inverse relationship. For example, with more strikeouts, fewer runs are scored on average. Coefficients are returned as part of multiple linear regression analysis in Excel and can be used to create formulas that evaluate existing data or help predict future events.

BUSINESS EXAMPLE: SPORT MARKETING ANALYSIS

Imagine a sport organization wants to see how different marketing strategies affect their ticket sales. The dependent (response) variable is the ticket sales volume, and the independent

An Introduction to Applied Sport Analytics **https://doi.org/10.1016/B978-0-443-49068-2.00006-9**
© 2027 Elsevier Inc. All rights are reserved, including those for text and data mining, AI training, and similar technologies.

variables could be advertising expenditure, social media activity, and price promotions. After running a multiple linear regression analysis, we get the following results:

- Advertising expenditure has a positive coefficient and a *P*-value of .0087999. This means spending more on ads may increase ticket sales. Due to the low *P*-value, the result is considered to be statistically significant since it is below .05.
- Social media activity has a positive coefficient and a *P*-value of .003241227. This suggests spending more on social media activity may increase ticket sales. The *P*-value is also considered to be statistically significant.
- Price promotions have a positive coefficient and a *P*-value of .009806784. This indicates that offering discounts or limited-time deals may increase ticket sales. The *P*-value is also considered to be statistically significant.

The organization can now use the output of the multiple linear regression analysis to guide decisions on spending for advertising, social media, and price promotions to increase sales.

COLLINEARITY AND MULTICOLLINEARITY

Collinearity or multicollinearity occurs when predictors in a regression model are moderately or highly correlated with one another (Field, 2013). This means that some independent variables provide overlapping or redundant information about the dependent variable, which can make the model's estimates less reliable. Let's say we're analyzing baseball performance using multiple linear regression. Our independent variables are:

- Hits
- Singles
- Doubles
- Triples
- Home Runs

In this case, Hits is a broad category that includes Singles, Doubles, Triples, and Home Runs. Including Hits along with these other specific types of hits can cause multicollinearity issues because they are all closely related. This redundancy can lead to difficulties in determining the individual effect of each variable on the dependent variable. For this reason, it's important to understand your subject (know the sport) and to also give some thought to the independent variables that are being used. It would be a good idea to leave the variable "hits" out of this model.

MULTIPLE LINEAR REGRESSION AND BASEBALL

Let's run an analysis of Major League Baseball teams using multiple linear regression. Creating a multiple linear regression analysis in Excel requires the Data Analysis add-in. See Chapter 4 for instructions on activating the add-in to Excel.

1. Download the file "MLB Team Data.xlsx" from the book's website. Open this file in Excel.
2. Let's take a look at how power hitting (doubles, triples, and home runs) impacts total runs scored by a team. From the Data tab, select Data Analysis.

3. For Analysis Tools, select Regression and click OK (Fig. 6.1).

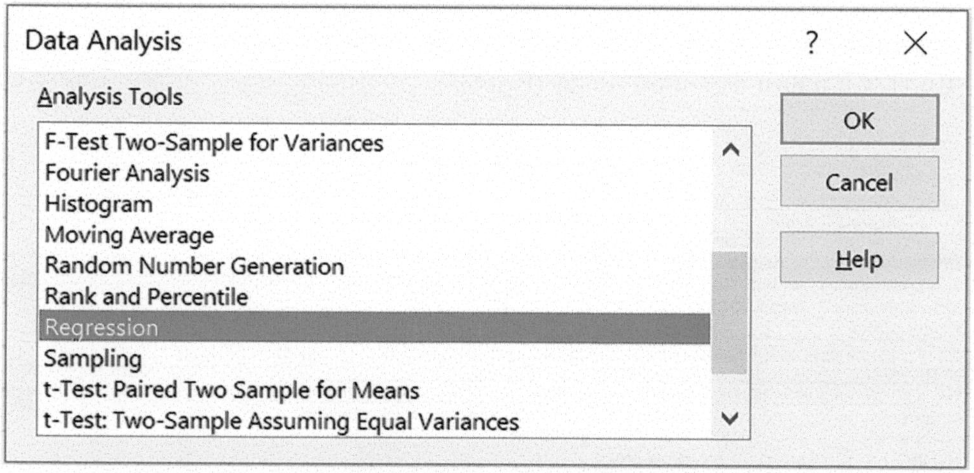

Figure 6.1 ***Regression.*** Selection of Regression from Analysis Tools.

4. For Input Y Range, drag the cursor over all of column B, where the R stands for the number of runs a team scored that season. This is our dependent variable.
5. For Input X Range, drag the cursor over all of the 2B, 3B, and HR data (Fig. 6.2).

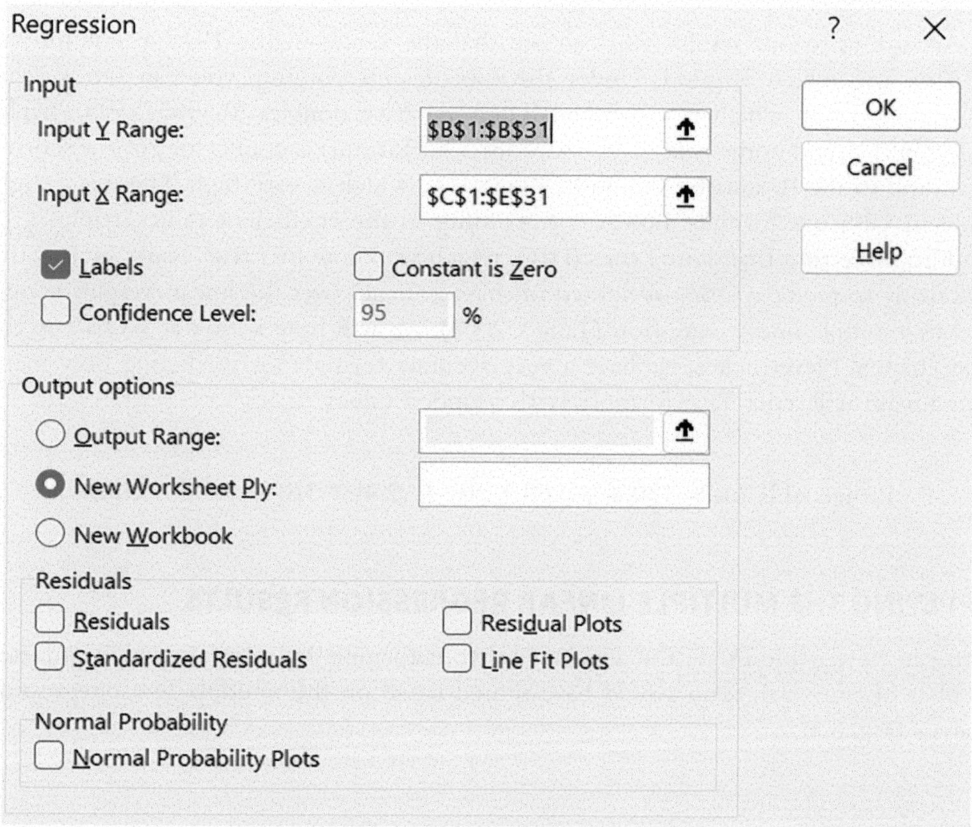

Figure 6.2 ***Labeling.*** How to label independent variables in the multiple linear regression model.

Note: If you also dragged the cursor over the column headings, click the box next to Labels.

6. Click the OK button.
7. Excel will return regression results (Fig. 6.3).

	A	B	C	D	E	F	G	H	I
1	SUMMARY OUTPUT								
2									
3	*Regression Statistics*								
4	Multiple R	0.871686896							
5	R Square	0.759838044							
6	Adjusted R Square	0.732127049							
7	Standard Error	39.05815972							
8	Observations	30							
9									
10	ANOVA								
11		*df*	*SS*	*MS*	*F*	*Significance F*			
12	Regression	3	125491.3308	41830.4436	27.42009254	3.26124E-08			
13	Residual	26	39664.03586	1525.539841					
14	Total	29	165155.3667						
15									
16		*Coefficients*	*Standard Error*	*t Stat*	*P-value*	*Lower 95%*	*Upper 95%*	*Lower 95.0%*	*Upper 95.0%*
17	Intercept	139.4020534	85.06642529	1.638743523	0.113313138	-35.45448798	314.2585949	-35.45448798	314.2585949
18	2B	0.596575505	0.353796008	1.686213215	0.103720045	-0.130662606	1.323813615	-0.130662606	1.323813615
19	3B	2.489368111	0.898406641	2.770870113	0.010186879	0.642666813	4.336069409	0.642666813	4.336069409
20	HR	1.978701738	0.287291322	6.887439983	2.59898E-07	1.388165968	2.569237508	1.388165968	2.569237508

Figure 6.3 ***Regression Results.*** Summary output of regression results.

According to our results, you can see that the values in the *P*-value column are all very low, which is good. Under the Coefficients column, you can see a value for our intercept, which is a constant. Then we have doubles (0.596575505), triples (2.48936811), and home runs (1.978701738). Our formula accounts for 76% of scoring according to the R-squared value (0.759838044), which is very high. There is a slight problem with our formula, however. According to the coefficient values, triples (3B) result in more runs than home runs (HR), which we know isn't true. Issues such as this lead many to prefer wOBA (weighted on-base average) over the linear weights generated by multiple linear regression (Tango, 2008). We will have a look at wOBA in the next chapter. Nevertheless, we have a very accurate formula for predicting how many runs a team will score. That formula, with rounded values, is:

$$\text{Projected Runs} = 139.4 + (.60 \star 2B) + (2.49 \star 3B) + (1.98 \star HR)$$

APPLYING THE MULTIPLE LINEAR REGRESSION RESULTS

You can now demonstrate the accuracy of your formula by calculating the projected number of runs each team would have scored based on the multiple linear regression analysis (Fig. 6.4).

SUM =139.4+(0.6*C2)+(2.49*D2)+(1.98*E2)

	A	B	C	D	E	F
1	Team	R	2B	3B	HR	Projected Runs
2	Arizona	886	271	37	211	=139.4+(0.6*C2)+(2.49*D2)+(1.98*E2)
3	Oakland	643	240	19	196	
4	Atlanta	704	273	15	213	
5	Baltimore	786	262	33	235	
6	Boston	751	311	30	194	

Figure 6.4 ***Applying Multiple Linear Regression Results.*** How projected runs for MLB are being computed in Excel based on multiple linear regression analysis.

Once the data for Projected Runs column has been populated for every team, you can compare this data to the actual number of runs they scored in a scatter plot (see the previous chapter for specifics on how to create a scatter plot) (Fig. 6.5).

Figure 6.5 ***Scatter Chart.*** Scatter chart of actual runs versus projected runs.

We can now start to see the predictive power of multiple linear regression, as power-hitting players are traded or acquired in free agency each season. Analysts can project the impact on runs scored and, ultimately, wins based on expected performance. What other offensive categories would you include to try to predict how many runs a team will score over the course of a season? Will they be statistically significant?

Mini Project

1. Download the dataset "Ticket Sales Data.xlsx" from the book's website. The file includes columns for Ticket Sales, Advertising Expenditure, Social Media Activity, and Price Promotions.
2. In Excel, run a multiple linear regression using Ticket Sales as the dependent variable and the other three factors (Advertising, Social Media, and Promotions) as independent variables.
3. Review the coefficients and *P*-values in your results. Which marketing factors appear to have the greatest impact on ticket sales? Are all statistically significant?
4. Write a short (200–300 word) summary advising a sport organization on how to allocate its marketing budget based on your findings. Explain which strategies seem most effective and how this data-driven approach could improve future ticket sales or fan engagement.

SUMMARY

Multiple linear regression is a valuable tool for understanding how different factors work together to influence an outcome. By looking at the coefficients, you can see the direction and strength of the relationships, and the *P*-values help you determine the reliability of those relationships. This method can be applied in various fields, from business and marketing to education and health, providing insights that can inform better decisions and strategies.

BEYOND THE BOX SCORE

1. Multiple linear regression allows us to examine how several variables together affect an outcome. How does this deepen your understanding of performance or business strategy compared to using a single variable alone?
2. In the baseball example, triples were shown to have a larger coefficient than home runs, even though we know home runs generally contribute more to scoring. What might this tell you about the limitations of regression analysis, and how should analysts handle results that don't align with common sense or domain knowledge?
3. Collinearity and multicollinearity can distort regression results. How can your understanding of the sport or business context help you decide which variables to include or exclude in your model?
4. The sport marketing example in this chapter showed statistically significant relationships between ticket sales and multiple forms of outreach. How can these insights inform decisions about where to allocate resources in a sport organization's marketing budget?
5. What part of the multiple linear regression process did you find most challenging or most rewarding? How did this activity shape your understanding of how regression can be used in sport analytics?

AI DISCLOSURE

During the preparation of this work, the author(s) used ChatGPT (OpenAI GPT-5) in order to provide editing and improve the readability of the document. After using this tool/service, the author(s) reviewed and edited the content as needed and take(s) full responsibility for the content of the publication.

REFERENCES

Field, A., 2013. Discovering Statistics Using IBM SPSS Statistics. 4th ed. Sage.
Lohr, S. L., 2023. Lies, damned lies, and statistics misquotations. SharonLohr.com. https://www.sharonlohr.com/blog/2023/10/31/statistics-misquotations/.
Tango T., 2008. Regression. schmegression. *InsideTheBook.com*. https://www.insidethebook.com/ee/index.php/site/regression_schmegression/.

CHAPTER 7

Expected Value Metrics

Key Takeaways

- Linear weights and wOBA help measure a player's true offensive value by giving different importance to singles, doubles, triples, and home runs.
- Run expectancy and run frequency show how likely teams are to score based on the number of outs and runners on base.
- Analytics can guide strategy, but managers must still balance data, experience, and instinct when making in-game decisions.

We use probability theory every day, as it provides a framework for dealing with the uncertainty. The nice thing about baseball is that all of the possible outcomes are known—it's not quite as messy as the real world. That makes the game an excellent playground for probability.

– Paul DePodesta, long-time MLB and NFL executive (Berger, 2016)

Baseball is a game rich in statistics, and over the years, advanced metrics have been developed to better understand and quantify player performance. This chapter explores linear weights, weighted on-base average (wOBA), run expectancy, run frequency, and probability. These metrics and theories, coupled with correlation (see Chapter 4) and regression (see Chapters 5 and 6), deepen insights into how baseball players contribute to their teams and help managers make informed decisions.

LINEAR WEIGHTS

Linear weights are values assigned to different events in a sport (like baseball singles, doubles, triples, and home runs) to quantify their impact on scoring. By assigning a specific value to each type of event, linear weights help to measure a player's offensive contribution more accurately than traditional statistics, like batting average or slugging percentage.

For example, in baseball a home run has a greater impact on the game than a single, and linear weights account for these differences. Imagine a single is worth on average 0.888 runs, a double is worth 1.271 runs, a triple is worth 1.616 runs, and a home run is worth 2.101 runs according to a linear model. These values show how much each event contributes to the overall scoring potential of the team on average. Note that we

An Introduction to Applied Sport Analytics **https://doi.org/10.1016/B978-0-443-49068-2.00007-0**
© 2027 Elsevier Inc. All rights are reserved, including those for text and data mining, AI training, and similar technologies.

know that an actual home run never contributes 2.101 runs, but since we are dealing with averages, the average does not have to be an actual outcome of an at-bat.

WEIGHTED ON-BASE AVERAGE

Weighted on-base average (wOBA) is a comprehensive statistic that measures a player's overall offensive value by assigning different weights to different outcomes. Developed by Tom Tango, wOBA is widely regarded as the best single metric for measuring a hitter's production (Law, 2018). The weights in the formula are derived from the average run value of different events. An example of the formula for wOBA is:

$$\frac{(0.690 \times \text{uBB} + 0.722 \times \text{HBP} + 0.888 \times 1\text{B} + 1.271 \times 2\text{B} + 1.616 \times 3\text{B} + 2.101 \times \text{HR})}{(\text{AB} + \text{uBB} + \text{SF} + \text{HBP})}$$

The formula might seem complicated or intimidating, but it is simply adding weighted values to a variety of events. The abbreviations in this formula are:

- uBB is unintentional walk and is worth 0.690 runs on average.
- HBP is hit by pitch and is worth 0.722 runs on average.
- 1B is singles and these are worth 0.888 runs on average.
- 2B is doubles and these are worth 1.271 runs on average.
- 3B is triples and these are worth 1.616 runs on average.
- HR is home runs and these are worth 2.101 runs on average.
- AB is at-bats.
- SF is sacrifice flies.

You can see how a formula like wOBA, which accounts for a variety of counting stats, presents a more complete picture of a batter than does batting average or other statistics. It should be intuitive that a home run is better than a single for a hitter and, hence, the coefficient of HR (2.101) is greater than the coefficient of 1B (0.888). wOBA combines the value of getting on base with the value of hitting for power, making it a comprehensive measure of offensive performance.

RUN EXPECTANCY

Run expectancy is a metric that estimates the average number of runs a team can expect to score from a given base-out state to the end of the inning. It is a valuable tool for managers to assess the potential impact of different strategies during a game. Run expectancy is particularly useful early in the game when teams aim to score as many runs as possible. For example, with a runner on first base and no outs, run expectancy can help managers decide whether to swing away, sacrifice bunt, or attempt a steal. The answer using run expectancy is often to let the batter swing away.

Here's an example of a run expectancy chart. You can see on the left of the chart that there are eight potential base states, from bases empty to bases loaded. On the top of the chart, there are three potential out states. Thus there are a total of 27 different states within the sport of baseball or softball. There isn't a column for 3 outs because we know that the values for that column would be zero, since no runs are scored with 3 outs (Fig. 7.1).

Bases	0 outs	1 out	2 outs
Empty	0.495	0.271	0.104
1st	0.905	0.538	0.230
2nd	1.050	0.654	0.317
3rd	1.325	0.911	0.341
1st&2nd	1.504	0.952	0.452
1st&3rd	1.740	1.184	0.472
2nd&3rd	1.886	1.300	0.559
Loaded	2.373	1.649	0.752

Figure 7.1 ***Run Expectancy Chart.*** Run expectancy chart for baseball.

The chart can be very valuable when runners get on base. We can see the value of getting a leadoff hitter on base. By getting a runner on first base with no outs, the average number of runs scored for this team jumps from 0.495 to 0.905. This is an increase of 0.410, which is a tremendous increase in run scoring that emphasizes the importance of getting a leadoff hitter on base.

The chart also demonstrates the futility of sacrifice bunting as a strategy (in general). In no instance does giving up an out to advance a runner increase the total number of runs a team will score over the course of a season. That's not to say that there is no reason to perform a sacrifice bunt (see the "Run Frequency" section), but it certainly raises a concern about managers who use this tactic on a consistent basis.

RUN EXPECTANCY AND PROBABILITY THEORY

We can use run expectancy to estimate the necessary success rates for baseball or softball strategies such as bunting and stealing bases. To do this, we'll take advantage of probability theory, which is a field in statistics that is used to help predict the likelihood of different events occurring (Helwig, 2020). Using our previous run expectancy chart, let's say our leadoff hitter gets on base and we want them to steal second base. We know that we average almost 0.905 runs if we leave that runner at first base. If that base runner successfully steals second base, our run expectancy goes up to 1.050. However, if the base runner is thrown out, we will have the bases empty and one out, where our run expectancy goes down to 0.271.

We can calculate the break-even point for stealing bases by using this probability formula:

$$(\text{Current State} - \text{Failure})/(\text{Success} - \text{Failure})$$

Let's do the math. Our current state at first base with no outs is 0.905 in run expectancy. Failure in this situation means the runner is thrown out stealing, so in that case, there would be no runners on base and one out, which is 0.271. Success means the runner steals second and there will be no outs, which is 1.050. Here's our formula with these values:

$$\text{Break even} = (0.905 - 0.271)/(1.050 - 0.271)$$

This works out to 0.814 or 81.4%. For stolen bases to be successful as a strategy to increase our overall run scoring, a runner at first with no outs must be above the break-even point and steal successfully about 82% of the time on average or better. Due to rule changes in Major League Baseball that favored base stealers in 2024, players were successfully stealing at about an 80% rate. This was the highest success rate in history. In Division I women's college softball, the success rate was 82.71% in 2025.

Let's do this same math for a runner on first but with one out and two outs. Our current state at first base with one out is 0.538 in run expectancy. Failure in this situation means the runner is thrown out stealing, so in that case, there would be no runners on base and two outs, which is 0.104. Success means the runner steals second and there will be one out, which is 0.654. Here's our formula with these values:

- Break even with one out = (0.538–0.104)/(0.654–0.104). This works out to be about 0.789 or 78.9%.
- Break even with two outs = (0.23–0)/(0.317–0). This means the break-even point is around 72.6%.

Note: With two outs, the failure value is zero because it ends the inning. In other words, no runs can score with three outs.

But what about stealing third? Here are the necessary success rates for stealing third base:

- 0 outs—73.9%
- 1 out—68.2%
- 2 outs—93.0%

Interestingly, performing the calculations for a runner at second base supports the baseball axiom that a base runner should never make the first or third out at third base. Therefore, if there's a situation to steal third base, it's with one out since the necessary success rate is lower. Also, only an elite base runner should consider an attempt to steal third base with two outs since the necessary success rate (93.0%) is so high.

These same calculations can be applied to bunting as well. However, according to the run expectancy chart, there is no situation in which a sacrifice bunt increases the overall number of runs scored. As a result, bunting tends to decrease run production if used regularly over the course of a season. In Chapter 1, we referred to the work of Dr. George Lindsey, who demonstrated that the most important run in a game is the one that ties the score or puts a team ahead by a single run late in the game. Thus there is a place for sacrifice bunting, and this strategy can best be illustrated through run frequency charts.

RUN FREQUENCY

There is more to the story than just run expectancy, which can be used when teams are using a strategy to score as many runs in an inning as possible. But there are also situations where teams are playing for just one run, which often happens late in a close game. This is where run frequency charts come in handy. Run frequency charts provide the probability of one run scoring from a specific base-out state during the remainder of the inning. This is an important difference between the two charts. Run expectancy estimates the total number of runs scored from a given base and out state while run frequency provides the likelihood (percentage) of scoring just one run from a given base and out state. Run frequency helps managers make decisions about small-ball strategies like bunting and stealing, especially in late-game situations where one run can determine the outcome of a game.

If a team is trailing by one run in the late innings with a runner on second base and no outs, a run frequency chart can show the likelihood that the runner will score. This information can guide the manager's decision to bunt the runner to third, attempt a steal, or swing away.

Here is an example of a run frequency chart (Fig. 7.2).

Bases	0 outs	1 out	2 outs
Empty	0.261	0.154	0.065
1st	0.415	0.264	0.119
2nd	0.575	0.399	0.223
3rd	0.841	0.654	0.251
1st&2nd	0.622	0.434	0.233
1st&3rd	0.847	0.661	0.256
2nd&3rd	0.847	0.661	0.256
Loaded	0.879	0.713	0.318

Figure 7.2 ***Run Frequency Chart.*** Run frequency chart for baseball.

With a runner on first base and zero outs, this team will score one run 41.5% of the time (shown as 0.415 above). That percentage decreases slightly (39.9%) if the runner is sacrifice bunted to second base. However, bunting for one run does make sense in some instances. For example, with a runner on second base and no outs, bunting the runner to third increases the likelihood of scoring one run from 57.5% to 65.4%. The percentage of scoring one run also increases when successfully sacrificing runners at first and second with no outs. So small ball does have a place statistically, which is typically near the end of a close game.

THE 2004 ALCS: BOSTON RED SOX VERSUS NEW YORK YANKEES—GAME 4

A classic moment that illustrates the real-time application of run expectancy and run frequency occurred during Game 4 of the 2004 ALCS between the Boston Red Sox and New York Yankees. Here are the run expectancy and run frequency charts for the Red Sox in 2004, which like previous charts were generated using Tom Tango's Markov calculator found at https://tangotiger.net/markov.html (Tango, n.d.) (Fig. 7.3).

2004 Boston Red Sox Run Expectancy

Bases	0 outs	1 out	2 outs
xxx	0.667	0.361	0.137
1xx	1.140	0.673	0.283
x2x	1.284	0.805	0.390
xx3	1.514	1.030	0.420
12x	1.805	1.154	0.548
1x3	1.995	1.350	0.573
x23	2.138	1.482	0.680
123	2.694	1.887	0.904

2004 Boston Red Sox Run Frequency

Bases	0 outs	1 out	2 outs
xxx	0.326	0.201	0.088
1xx	0.473	0.312	0.146
x2x	0.617	0.444	0.253
xx3	0.665	0.481	0.265
12x	0.848	0.669	0.283
1x3	0.855	0.678	0.290
x23	0.855	0.678	0.290
123	0.889	0.733	0.356

Figure 7.3 ***Run Expectancy and Run Frequency Charts.*** Run expectancy and run frequency charts for 2004 Boston Red Sox.

In the bottom of the ninth inning, the Red Sox were trailing 4–3 and facing elimination down 3 games to 0. Kevin Millar walked to lead off the inning. Red Sox manager Terry Francona made the decision to insert Dave Roberts, one of the fastest base runners in the league, as a pinch runner at first base with no outs.

This decision opened several options, including:

- Should Francona let the batter swing away, hoping to score two runs to potentially win the game outright? In that case, he'd be using the run expectancy chart. With a runner on first base and no outs, the Red Sox averaged 1.140 runs. On average, this is at least enough to tie the game.
- Should Francona call for a bunt, sacrificing the out in exchange for better run frequency odds of scoring the tying run? With a sacrifice bunt, Francona would be lowering the likelihood of scoring at least one run from 47.3% to 44.4%.
- Should Francona let Roberts attempt to steal second, risking an out for putting a speedy runner in scoring position where run frequency increases significantly? If successful, the chance of scoring from second jumps from 47.3% to 61.7%. The downside would be, if Roberts is thrown out stealing, the likelihood of scoring one run dropped to 31.2%.

Run expectancy models might suggest playing for a big inning but run frequency charts, especially for a successful base stealer like Roberts, support playing for just one run to tie the game and survive. Other variables made the decision even more complex:

- How good was Jorge Posada, the Yankees catcher, at throwing out runners?
- How good was the pitcher? In this case, it was Mariano Rivera, the greatest relief pitcher of all time.
- Was Rivera slow to the plate?
- What was the ability of the upcoming batters?

All of these factors had to be evaluated in seconds under immense pressure, which highlights how analytics, experience, and instinct all intersect in high-stakes moments.

Roberts stole second base successfully. Moments later, after first attempting a sacrifice bunt, Bill Mueller singled to center, tying the game. The Red Sox went on to win in extra innings and eventually won the series. Boston broke the "Curse of the Bambino" and became the first team in professional baseball to rally in a series from a 3–0 deficit.

HINDSIGHT BIAS

Had Dave Roberts been thrown out attempting to steal second base, critics would likely have questioned Francona's decision. This brings up the concept of hindsight bias, also known in sports as Monday morning quarterbacking. Hindsight bias is the tendency to judge decisions based on their outcomes rather than the information available at the time.

In this case, success made Francona look brilliant, but a failed steal would have likely been viewed as a blunder. However, the data supported the decision. Analytics can inform decisions, but they don't guarantee outcomes. Evaluating strategy requires understanding the process behind the choice, not just whether it worked.

SUMMARY

Baseball metrics like wOBA, run expectancy, and run frequency provide deeper insights into player performance and in-game strategies. By understanding these metrics, players, coaches, and fans can appreciate the game on a more sophisticated level, recognizing previously overlooked factors that contribute to winning and losing. Probabilities also highlight the difficulty of in-game decision-making, where split-second decisions are made by managers that are often criticized or celebrated in hindsight.

For a deep dive into the value of these statistics, consider *The Book: Playing the Percentages in Professional Baseball* by Tom Tango, Mitchel Lichtman, and Andrew Dolphin (Tango, Lichtman, and Dolphin, 2007). This is a seminal book in baseball analytics.

BEYOND THE BOX SCORE

1. Paul DePodesta described baseball as a "playground for probability." How can run expectancy be used to show the power of applying probability to evaluate performance and strategy?
2. Run expectancy suggests sacrifice bunting is rarely a good choice, yet run frequency shows it can make sense in some late-game situations. How can understanding both charts help managers make smarter in-game decisions?
3. In the 2004 ALCS, Red Sox manager Terry Francona made a high-stakes decision to pinch run and attempt a steal. What does this moment illustrate about the role of analytics, probability, and instinct in sport decision-making?
4. The break-even analysis for base stealing uses a simple formula but provides powerful insights. How might you apply a similar approach to evaluate other strategies in sport, such as pulling a goalie in hockey or fouling late in a basketball game?
5. Hindsight bias often impacts how decisions are judged in sport. Why is it important to separate decision quality from outcome, and how can analytics help support or defend decisions even when they don't work out?

Looking ahead: Now that you've learned the fundamentals of data analysis (such as the Pythagorean Theorem, correlation, and regression), you are ready to bring those insights to life through data visualization. In the next section, you'll learn how to use visualizations to make information easier to understand and share. You'll see how charts, graphs, and dashboards can reveal patterns that numbers alone might miss. By turning data into visuals, you can tell clearer stories and make better decisions in sport analytics.

Mini Project

1. Download the dataset "Texas Softball Run Expectancy 2025.xlsx" from the book's website. The file includes run expectancy values for all 24 base-out states for the 2025 D-I women's college softball national champions, the University of Texas Longhorns.
2. Use the run expectancy formula to calculate the break-even point for a stolen base attempt with a runner on first and no outs.
   Use these values from the chart:
   - Current state (runner on first, 0 outs): 1.656
   - Success (runner on second, 0 outs): 1.787
   - Failure (bases empty, 1 out): 0.586
   - Apply the probability formula:
   - Break even = (Current – Failure)/(Success – Failure)
3. After calculating the break-even percentage, identify two other base-out states where a steal might be considered, and calculate those break-even points using the same approach.
4. How can probability and expected value guide decisions? In another sport other than baseball or softball, when might intuition or game context override the data?

AI DISCLOSURE

During the preparation of this work, the author(s) used ChatGPT (OpenAI GPT-5) in order to provide editing and improve the readability of the document. After using this tool/service, the author(s) reviewed and edited the content as needed and take(s) full responsibility for the content of the publication.

REFERENCES

Berger, K., 2016. Revisiting Moneyball with Paul DePodesta, Nautilus, 25 August. Available at: https://nautil.us/revisiting-moneyball-with-paul-depodesta-236084/ (accessed 16.07.26.).

Helwig, N. E., 2020. Introduction to probability theory [Lecture notes]. University of Minnesota. Available at: https://users.stat.umn.edu/~helwig/notes/ProbabilityTheory.pdf (accessed 16.07.26.).

Law, K., 2018. Smart Baseball: the story behind the old stats that are ruining the game, the new ones that are running it, and the right way to think about baseball. 1st paperback edn. William Morrow.

Tango, T., n.d. Markov run estimator, The Book blog. Available at: https://tangotiger.net/markov.html (accessed 11.06.25.).

Tango, T. M., Lichtman, M. G. and Dolphin, A. E., 2007. The Book: playing the percentages in baseball. Potomac Books.

SECTION III

Visualizing the Game

This section introduces the fundamentals of data visualization. It begins with the core principles of effective visual communication and then builds toward using Power BI and Tableau to create interactive dashboards and uncover sport-specific insights. Each chapter offers a beginner-friendly approach designed to help readers gain confidence in presenting data clearly and transforming raw sport data into compelling stories that support better decision-making.

CHAPTER 8

Data Visualization and Communicating Insights

Key Takeaways

- Data visualization helps turn complex numbers into clear stories, making it easier to see patterns and make informed decisions.
- Effective visualizations help your audience understand information quickly and accurately.
- Combining visuals with storytelling makes data more engaging and memorable.

The way in which we relate to information depends on how it is designed.

– David McCandless, InformationIsBeautiful.net

(The influence of visual information: Data visualisation, 2021)

Data visualization transforms numbers into insights, making complex data accessible and actionable (Olowe et al., 2024). As humans, it is typically easier to understand information presented through pictures or graphs than it is through numbers. Therefore it is important to give thought to telling a story using data visualization. Effective data visualizations emphasize conclusions rather than overwhelming details (Knaflic, 2015). The goal is to highlight key insights and tell a story using clear, accurate visuals. Elements like color and shading can guide attention to crucial points, making the data both memorable and meaningful. Imagine a coach analyzing thousands of data points on player performance. While raw numbers tell a story, a well-designed heat map instantly reveals strengths, weaknesses, and trends. A basketball shooting heat map (see the image below) can instantly convey a player's shot selection trends from different areas on the court and may contribute to the decision-making process of how to guard a particular player (Fig. 8.1).

An Introduction to Applied Sport Analytics **https://doi.org/10.1016/B978-0-443-49068-2.00008-2**
© 2027 Elsevier Inc. All rights are reserved, including those for text and data mining, AI training, and similar technologies.

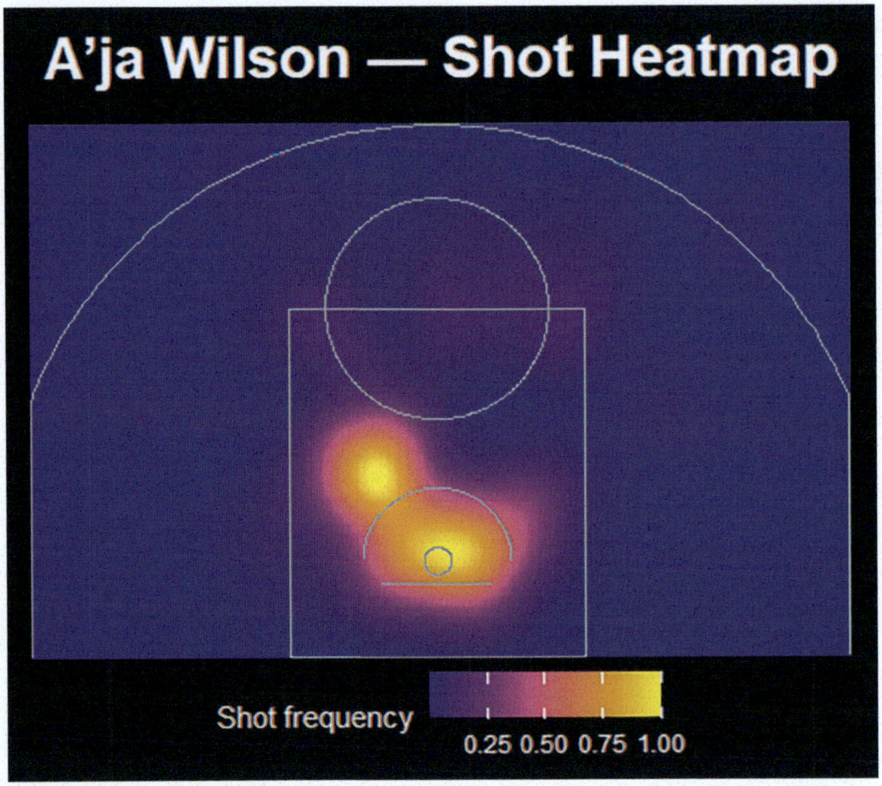

Figure 8.1 ***A'ja Wilson Shooting Heatmap.*** A'ja Wilson Shooting Heatmap showing location of shot selection.

Impactful visualizations possess several key qualities: they must be truthful, based on thorough and objective research, and functional, allowing users to act upon the information. They should also be visually appealing, drawing attention through well-designed and aesthetically pleasing displays, and insightful, providing information such as trends and inferences that are not immediately obvious. Furthermore, they should be enlightening, highlighting evidence, and telling a story in a way that is easy to understand.

Beyond communication, data visualization also supports analysis. A visualization can confirm assumptions, as seen in scatterplots that reveal correlations (Rensink, 2017). It can also be educational, such as with training sessions to report data simply or to develop insights within a limited time. Moreover, data visualization encourages exploration, allowing researchers to formulate new hypotheses (Heer and Shneiderman, 2012).

Incorporating predictive analytics into visualizations can provide foresight into future trends. For example, incorporating a trendline to show the correlation between team win percentage and attendance (see image below) helps identify how performance may influence future fan turnout (Fig. 8.2).

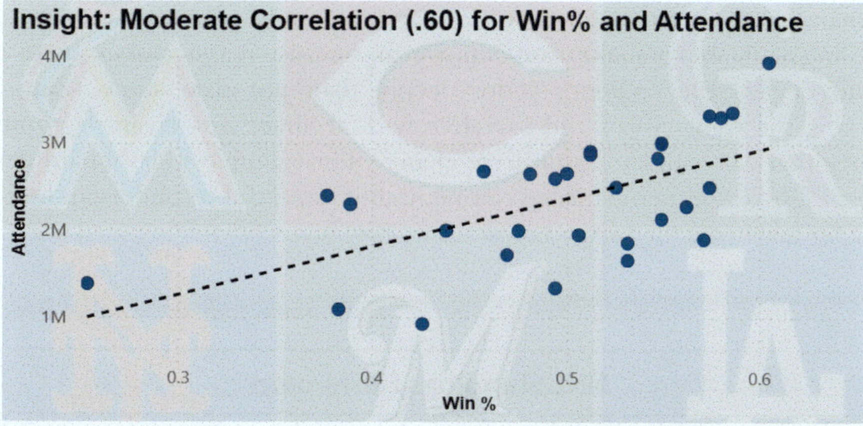

Figure 8.2 ***Win % and Attendance.*** Correlation of Win % and Attendance with a trendline.

Charts, as shown below, can also reveal that while some high-payroll teams achieve strong win totals, others underperform. Conversely, several low-payroll teams exceeded expectations. This chart demonstrates that while higher spending can be a significant advantage, it does not always guarantee success (Fig. 8.3).

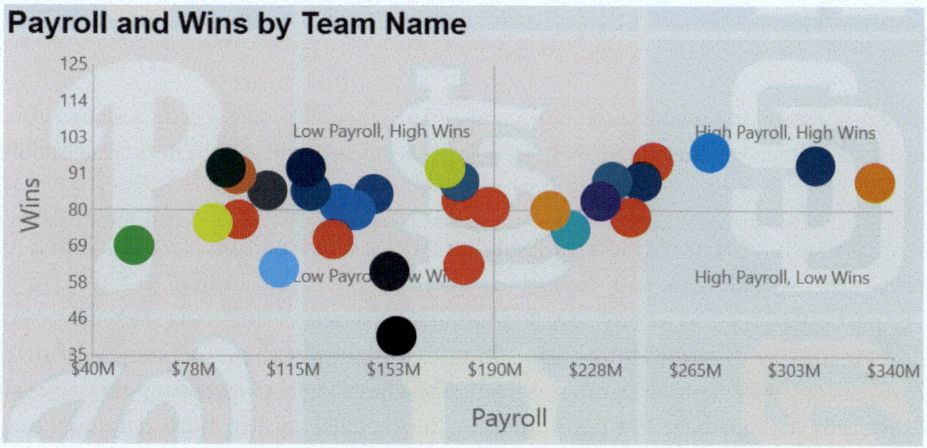

Figure 8.3 ***Payroll and Wins by Team.*** Payroll and wins by team using a dot plot.

Understanding the audience is essential in creating effective visualizations. One must consider the audience, their expectations, and whether the graphics should be exploratory or educational. Risks in data visualization include ensuring the quality of the underlying data, providing sufficient context to make data visuals approachable, and avoiding biases such as undue influence and cherry-picking data.

Humans have been telling stories for thousands of years (Dykes, 2019). Storytelling with data visualization is powerful because it can change brain chemistry and engage the audience more deeply than just data alone (Zak, 2015). Combining the right visuals and narrative with accurate data creates a compelling data story that can influence and drive change. For example, a dashboard detailing each team's in-season performance can be highly impactful to understanding sports at a deeper level (see the image below) (Fig. 8.4).

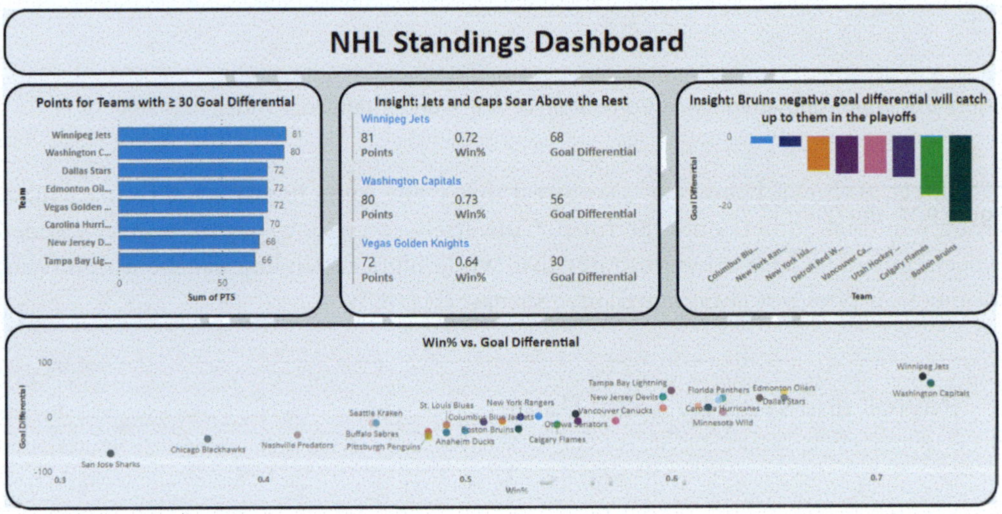

Figure 8.4 ***NHL Standings Dashboard.*** NHL Standings Dashboard with various data visualizations.

Key elements of data visualization include data, narrative, and visuals. When these elements are merged, they can engage or even entertain an audience. This combination is particularly effective in sports marketing, where visualizing fan engagement trends over time with compelling storytelling can highlight the effectiveness of marketing campaigns. Data visualization not only facilitates data storytelling but also aligns with different data analysis strategies, making it a crucial tool in both data analysis and communication.

Interactive data visualizations allow users to explore data in more depth. In sports, interactive dashboards can display real-time statistics and player performance metrics, providing immediate insights during a game. Visualizations are excellent for comparing different data points. For instance, the scatter chart shown below compares batter salary to their average weighted on-base average (wOBA) (Fig. 8.5).

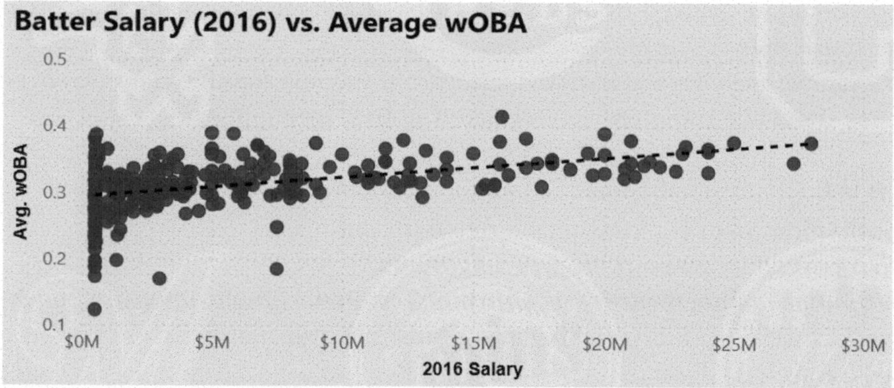

Figure 8.5 ***Batter Salary (2016) Versus Average wOBA.*** Batter salary (2016) versus average wOBA scatterplot with trendline.

Providing data visualizations that are accessible to all users, including those with disabilities, is important. This includes using high-contrast colors, providing text descriptions with insights, and ensuring that visualizations are readable by screen readers. Avoid relying solely on color to convey meaning (e.g., use labels or patterns), use readable font sizes, and ensure keyboard navigation is supported when visualizations are interactive. Making visuals accessible both supports users with disabilities and improves usability for everyone.

Good data visualizations should be scalable, meaning they can handle large datasets without losing clarity or usability. This is particularly relevant in sports analytics, where large volumes of data are often analyzed. The ability to visualize real-time data can be crucial in sports.

FOUR IDEAS FOR IMPACTFUL DATA VISUALIZATIONS

When creating visualizations, the acronym I.D.E.A. can help.

Insightful (I)

- The best visualizations reveal patterns, trends, and takeaways that might not be immediately obvious.
- Instead of just presenting raw numbers, they help the viewer understand what the data means.
- For example, a box plot of athlete performance metrics can uncover hidden variations among players, offering deeper understanding for coaches and analysts.

Data-driven (D)

- Effective visualizations rely on thorough and objective research.

- Data should be sourced from reliable databases and analyzed with appropriate statistical methods.
- An example includes a scatterplot comparing training intensity, and player performance should be based on real, verified data (not assumptions).

Ethical (E)

- Visualizations must be accurate and free from distortion.
- Avoid misleading scales or exaggerating trends.
- For example, an improper y-axis scale can skew performance differences, misleading viewers. Likewise, cherry-picking data points can distort analysis and lead to biased conclusions.

Aesthetic (A)

- Aesthetically pleasing designs draw attention and improve comprehension.
- Effective use of color, contrast, and layout enhances clarity without overwhelming the audience.
- Heat maps, trendlines, and well-organized dashboards can make data more digestible while keeping viewers engaged.

STORYTELLING WITH DATA

Data visualization is more than just creating charts—it's about telling a story. In sport analytics, visuals help people see what the numbers really mean. A simple chart can show how a team is improving, which promotions bring fans to games, or where a budget is being spent. The goal is to make data easy to understand and useful in decision-making.

When you create a visual, start by asking, "What story am I trying to tell?" Every chart should answer a question. A line chart showing attendance over the season tells one story, while a bar chart comparing ticket sales by night tells another. The type of chart you choose should fit the message you want to share.

Also, think about who will see your visual. Coaches might want to see player performance trends. Athletic directors might focus on revenue or fan engagement. Fans might just want to see how their favorite player or the team is succeeding. The way you present your data should reflect your audience.

Keep it simple and clear. Too many colors, shapes, or labels can make your message hard to follow. Highlight the most important points so your audience quickly understands the takeaway. As you move into Power BI, you'll learn how to design visuals that help others see the story behind the data.

SUMMARY

By adhering to these principles, data visualizations go beyond mere charts and graphs; they become powerful storytelling tools that can engage audiences, reveal meaningful patterns, and drive informed decision-making in sport analytics and beyond. Mastering data visualization is more than a technical skill—it's an art. By blending data, visuals, and storytelling, analysts can transform raw numbers into powerful insights that drive decisions. Whether in sports, business, or academia, the ability to present data effectively will always be a game changer.

Looking ahead: In the next chapters, you'll learn how to bring data visualization and storytelling to life using Microsoft Power BI. You'll see how to create interactive visuals and reports, connect datasets, and design visuals that make complex sport data easy to explore and understand.

BEYOND THE BOX SCORE

1. David McCandless said, "The way in which we relate to information depends on how it is designed." After working with visualizations, how do you see design influencing the way people understand and act on sport analytics data?
2. The chapter introduced the I.D.E.A. framework for creating impactful data visualizations. Which of the four elements (Insightful, Data-driven, Ethical, and Aesthetic) do you think is most often overlooked, and why is it important?
3. Visualizations like heat maps or trendlines can tell powerful stories. How can storytelling with data change the way sport organizations communicate with players, fans, or stakeholders?
4. We saw examples where high payroll didn't guarantee success. How can data visualizations uncover patterns that raw numbers might not reveal, and why is this important in sport decision-making?
5. Interactive dashboards allow users to explore data on their own. What are the benefits and potential risks of giving users this kind of control, especially in high-stakes sport environments?

Mini Project

1. Find a sport-related infographic that uses data to tell a story. This could come from a league website, sports article, marketing campaign, or social media post.
2. Study the infographic and consider:
   - What story is it trying to tell?
   - What visuals (charts, colors, icons, and images) are used to communicate the message?
   - Is the data accurate, clearly presented, and free from bias or distortion?
3. Evaluate the infographic using the I.D.E.A. framework (Insightful, Data-driven, Ethical, Aesthetic).
4. Write a short (250–300 word) analysis describing:
   - The main message of the infographic.
   - What makes it effective (or ineffective).
   - How it could be improved to tell a clearer or more engaging story.

AI DISCLOSURE

During the preparation of this work, the author(s) used ChatGPT (OpenAI GPT-5) in order to provide editing and improve the readability of the document. After using this tool/service, the author(s) reviewed and edited the content as needed and take(s) full responsibility for the content of the publication.

REFERENCES

Dykes, B., 2019. Effective data storytelling: how to drive change with data, narrative, and visuals. Wiley.

Heer, J., Shneiderman, B., 2012. Interactive dynamics for visual analysis. Commun. ACM 55 (4), 45–54. https://doi.org/10.1145/2133806.2133821.

Knaflic, C. N., 2015. Storytelling with data: a data visualization guide for business professionals. John Wiley & Sons.

MediaElite, 2021. The influence of visual information: data visualisation, 12 February. Available at: http://mediaelite.org/2021/02/12/the-influence-of-visual-information-data-visualisation/ (accessed 16.07.26.).

Olowe, K. J., Edoh, N. L., Zouo, S. J. C., Olamijuwon, J., 2024. Conceptual review on the importance of data visualization tools for effective research communication. Int. J. Eng. Res. Dev., 20 (11), 1259–1268.

Rensink, R. A., 2017. The nature of correlation perception in scatterplots. Psychon. Bull. Rev. 24, 776–797.

Zak, P. J., 2015. Why inspiring stories make us react: the neuroscience of narrative. Cerebrum: The Dana Forum on Brain Science. Available at: https://www.ncbi.nlm.nih.gov/pmc/articles/PMC4327528/ (accessed 16.07.26.).

CHAPTER 9

Introducing Microsoft Power BI

Key Takeaways

- Microsoft Power BI helps turn data into interactive visuals that make patterns and trends easier to understand.
- Power BI connects to many data sources.
- Because it builds on familiar Excel skills and integrates with tools like Teams and PowerPoint, Power BI is an accessible next step for anyone learning sport analytics.

The greatest value of a picture is when it forces us to notice what we never expected to see.
– John Tukey (Tukey, 1977)

Microsoft Power BI is a business intelligence and data visualization tool that helps turn raw data into useful insights. With Power BI, users can connect to different data sources and create interactive charts, graphs, and dashboards. These visualizations make it easier to see patterns, track changes over time, and make informed decisions. Power BI is widely used across many industries because it combines powerful features with a user-friendly design. Power BI is currently the most popular business intelligence tool on the market, competing with Tableau and numerous other options (see the image below) (Panchal, 2024) (Fig. 9.1).

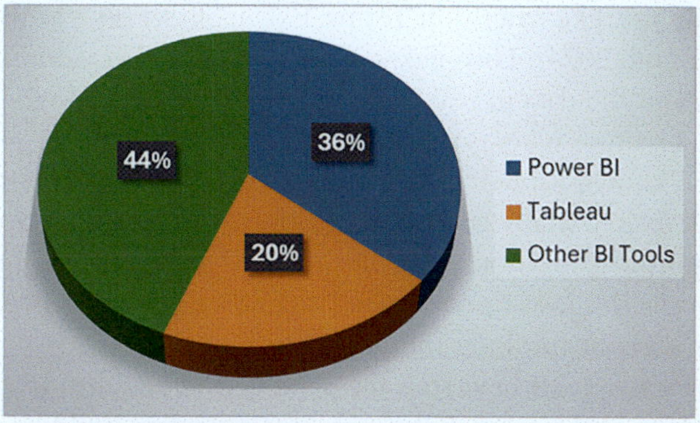

Figure 9.1 Pie chart illustrating the market share of business intelligence and data visualization tools.

An Introduction to Applied Sport Analytics https://doi.org/10.1016/B978-0-443-49068-2.00009-4
© 2027 Elsevier Inc. All rights are reserved, including those for text and data mining, AI training, and similar technologies.

In sport analytics, teams, coaches, analysts, and executives can use customized Power BI dashboards to track a variety of functions, including:

- Fan engagement and ticketing—analyzing attendance trends, ticket sales, and promotional campaign effectiveness.
- Financial performance—monitoring departmental budgets, revenue, and travel costs.
- Game strategy and opponent scouting—visualizing play outcomes and opponent tendencies.
- Injury tracking and return-to-play analysis—monitoring injuries, rehab timelines, and performance pre- and postinjury.
- Merchandise and concession sales—tracking revenue streams and inventory needs.
- Personnel—managing staffing roles, responsibilities, and turnover rates.
- Player performance—tracking key statistics, trends, and efficiency metrics.
- Prospects and recruits—evaluating scouting data and recruitment progress.
- Sponsorships—measuring sponsorship exposure and return on investment (ROI).

As shown in the following chart, Power BI is especially useful for answering practical questions both off and on the field.

Sport business questions and sport (on-field) questions

Sport business questions	Sport (on-field) questions
Which merchandise or promotions are most effective?	Which players are performing best?
What are the ticket sales trends by section or opponent?	How much of a training workload has a player had over the past month?
Where are fans experiencing delays or issues in the stadium?	How effective is a play or formation in different game situations?

Another advantage of Power BI is how well it supports collaboration. Instead of asking users to visit a separate website to view reports, Power BI can be integrated into tools people already use, like Microsoft Teams, Excel, SharePoint, PowerPoint, and even mobile apps (Microsoft, 2024). This makes it easier for coaches, administrators, and staff to view and talk about the data in real time. Having data show up where people already work helps create a culture where decisions are based on facts, not just opinions.

Power BI is also known for being affordable compared to other business intelligence tools (Knight et al., 2022). Many competing platforms with similar features cost several times more. For sport organizations with limited budgets, especially in college athletics or small-market teams, this lower price point can make a big difference. Teams can access powerful analytics without breaking the bank.

Finally, Power BI is a great next step for people who are already comfortable using Microsoft Excel. Many of the same skills (such as sorting data, building formulas, making charts, etc.) carry over to Power BI. In fact, Power BI was built using features from

Excel, such as Power Pivot and Power Query (Knight et al., 2022). If someone has built dashboards or reports in Excel before, they can often move those directly into Power BI and keep improving them. This makes the learning curve smoother and helps users get started more quickly.

The importance of Power BI in sport analytics lies in its ability to support data-driven decision-making. In today's competitive sports environment, having access to real-time, accurate information is critical. Power BI allows organizations to move beyond spreadsheets and create dynamic, interactive reports that keep everyone informed. To effectively use Power BI, it helps to understand its main components:

- Power BI Desktop—a computer program with a visual interface for building reports and dashboards.
- Power BI Service—a cloud-based platform where users can publish and share reports. This version of Power BI is what we will be using in the next chapter.
- Power BI Mobile—allows users to view and interact with reports on mobile devices.
- Power BI Report Server—hosts reports on-premises for organizations that need internal-only access.
- Power BI Gateway—connects on-site data sources to Power BI for real-time updates.

POWER BI SERVICE

In this hands-on training, you will learn how to use Power BI Service, the cloud-based version of Microsoft Power BI, to create an interactive data report. Step by step, you will build a report by selecting data fields, creating visuals, and customizing the layout using Power BI's easy-to-use tools. This guided activity is designed to help you become familiar with navigating the Power BI interface directly within your web browser.

CREATING A REPORT IN POWER BI SERVICE

Microsoft provides a free trial version of Power BI Service or what is sometimes called Power BI Online. To access Power BI Service in a web browser, go to app.powerbi.com. Create an account and sign in to Power BI Service.

Note: Access to Power BI Service requires a work or school email address. If you don't have one, you can sign up for a free 30-day Microsoft 365 trial, which includes a work account, at https://www.microsoft.com/en-us/microsoft-365/try. Alternatively, you can use Power BI Desktop, which does not require a work or school email, by downloading it at https://www.microsoft.com/en-us/download/details.aspx?id=58494. Many of the features in Power BI Service are similar to those in Power BI Desktop.

Once you have access, follow these steps.

1. Click [New report] (Fig. 9.2).

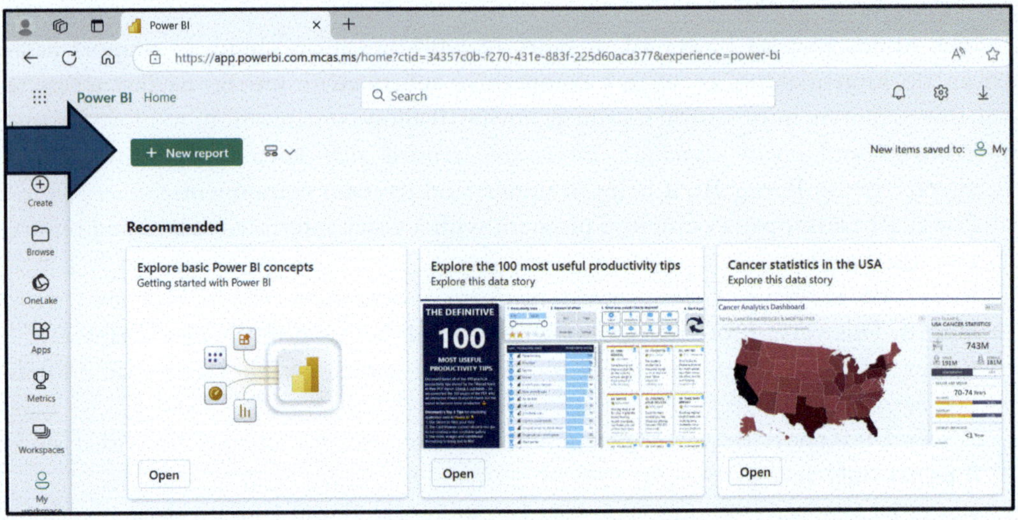

Figure 9.2 ***Power BI Opening Page.*** Power BI opening page with arrow pointing to new report.

2. Click Excel (Preview) (Fig. 9.3).

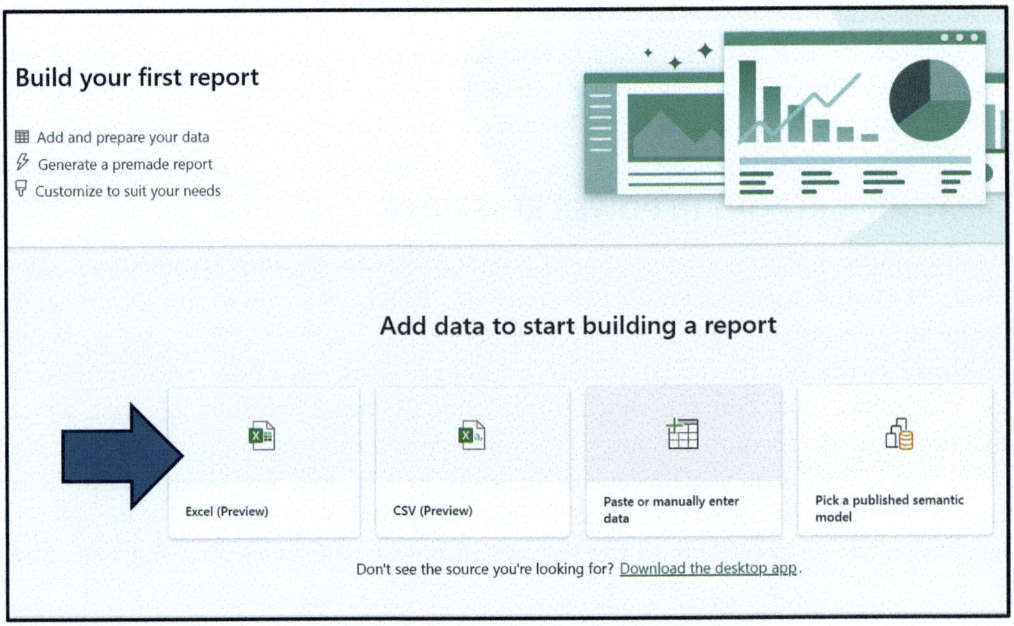

Figure 9.3 ***Excel (Preview).*** Arrow pointing where to click for Excel (Preview).

3. Click Upload File and [Browse] (Fig. 9.4).

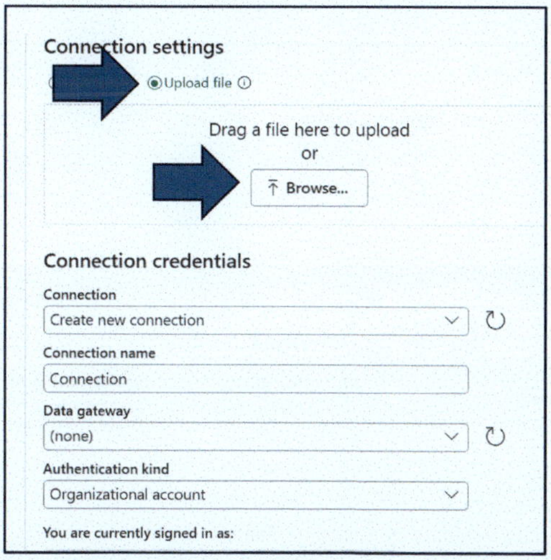

Figure 9.4 ***Click Upload File and Browse.*** Two arrows indicating where to click Upload File and Browse.

4. Download the file "WNBA 2024.xlsx" from the book's website. Navigate to the file and click [Open] (Fig. 9.5).

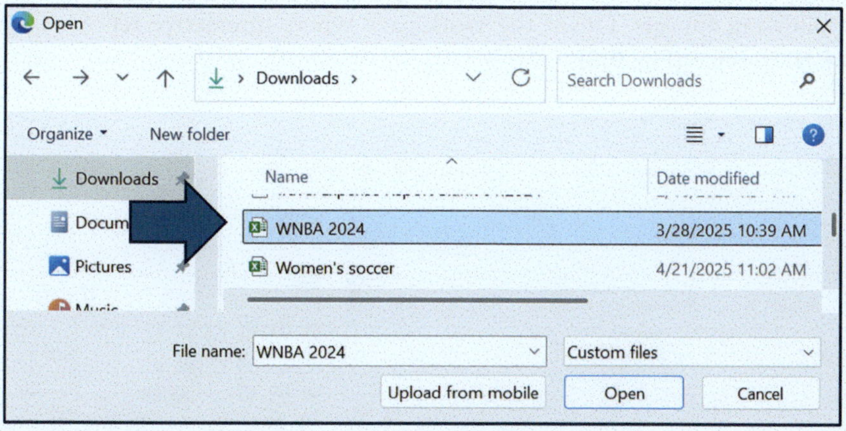

Figure 9.5 ***WNBA 2024 Data Set.*** Arrow pointing at WNBA 2024 Data Set.

5. Click [Next] (Fig. 9.6).

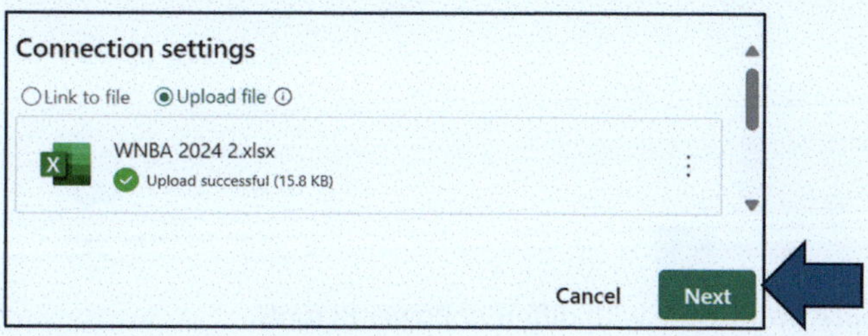

Figure 9.6 ***Next.*** Arrow pointing to Next button.

6. Click the box on the left next to WNBA. A preview of the data being imported from the spreadsheet displays on the right. Click [Create] (Fig. 9.7).

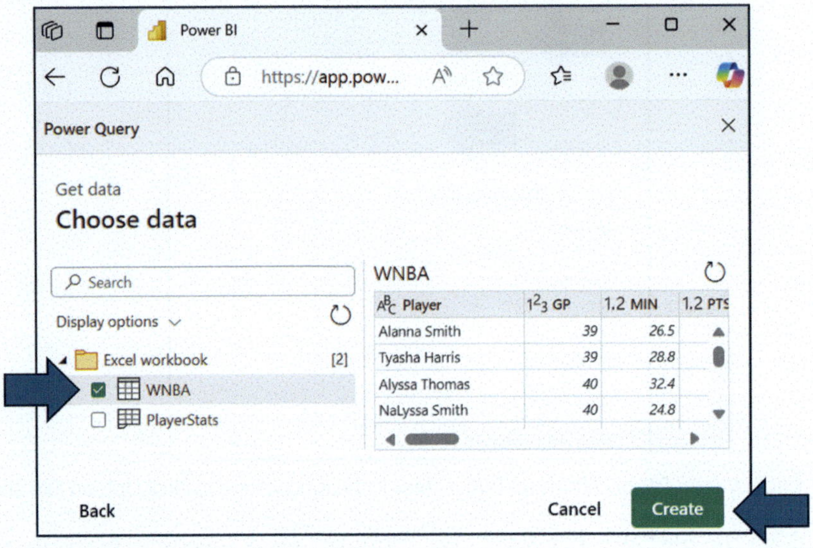

Figure 9.7 ***Choose Data.*** Arrow points toward selection of WNBA.

7. The following appears. This is the workspace where interactive reports are built in Power BI Service (Fig. 9.8).

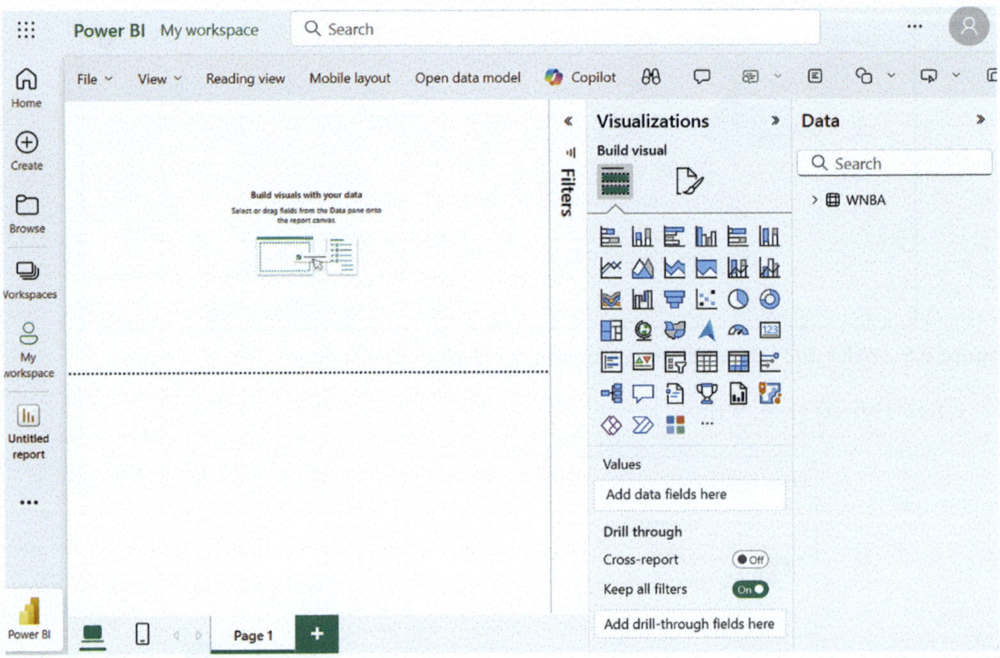

Figure 9.8 ***Workspace.*** Workspace with different Power BI tools.

8. On the right under Data, click WNBA. Each of the variables from the spreadsheet that was uploaded appears along with corresponding symbols. Put a check mark next to 3P%. The column chart on the left appears (Fig. 9.9).

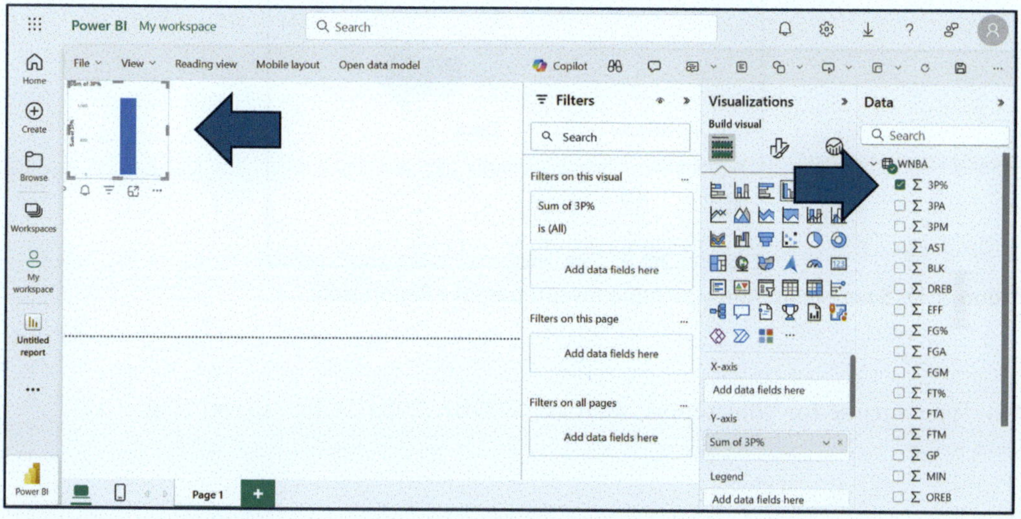

Figure 9.9 ***3P% Bar Chart.*** Arrow pointing toward 3P% Bar Chart.

9. Put a second check mark next to Player. The following column chart appears (Fig. 9.10).

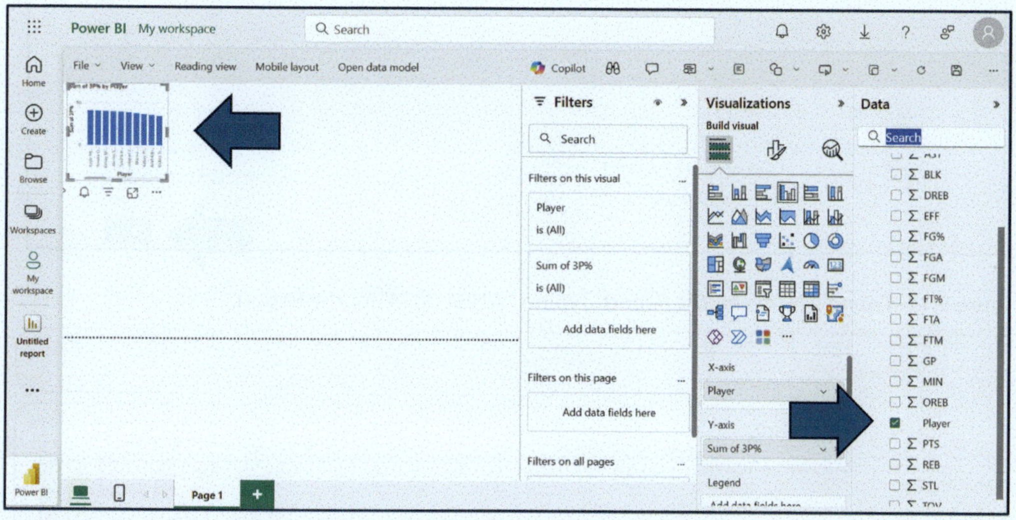

Figure 9.10 ***Bar Chart With Players.*** Arrow pointing toward result of a bar chart with players labeled.

10. Click the icon for Save this Report (Fig. 9.11).

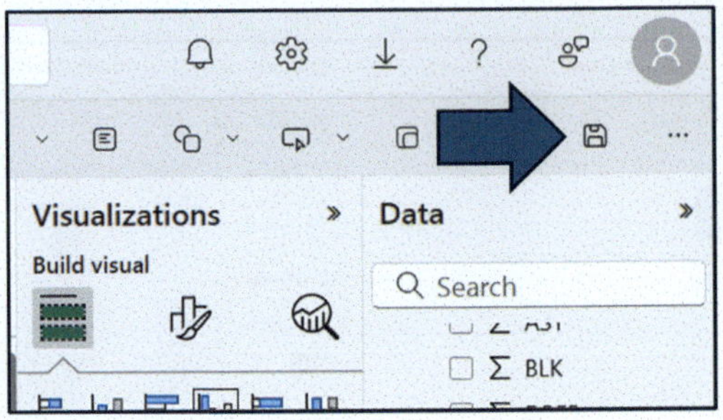

Figure 9.11 ***Save Report.*** Arrow pointing toward the Save Report icon.

11. Type a name for your report, such as "WNBA 2024." Click [Save] (Fig. 9.12).

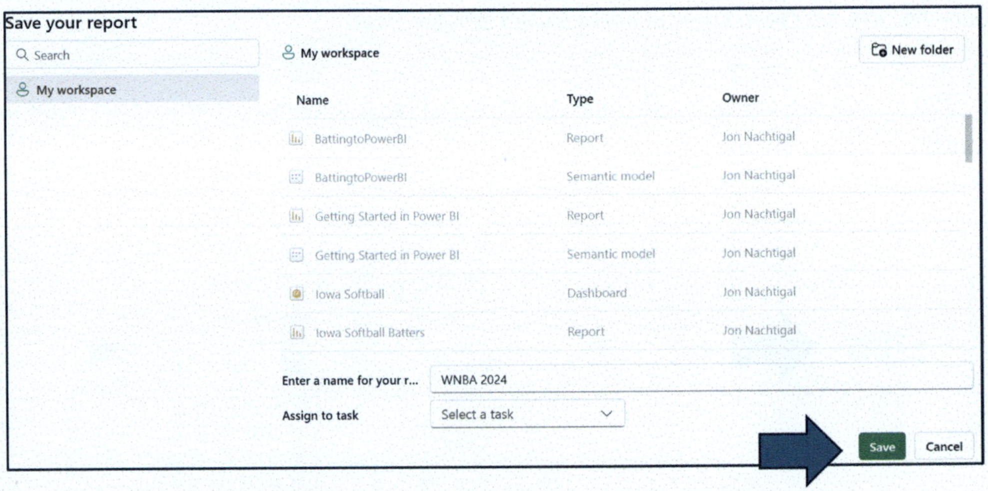

Figure 9.12 ***Typing a Name for the Report.*** Typing "WNBA 2024" for the Report.

12. The following page appears. Notice that the column headings on the right have disappeared. Click the Edit icon to activate the page (Fig. 9.13).

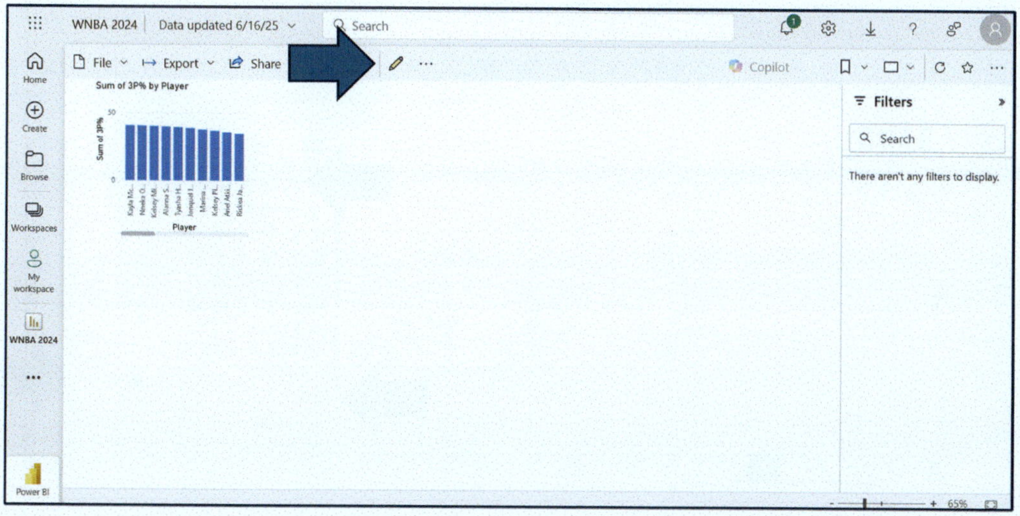

Figure 9.13 ***Sum of 3P% Bar Chart.*** Sum of 3P% Bar Chart with player names on the x-axis.

13. Left click the chart on the left, and then under Visualizations, click each of the chart types in the first column (Fig. 9.14).

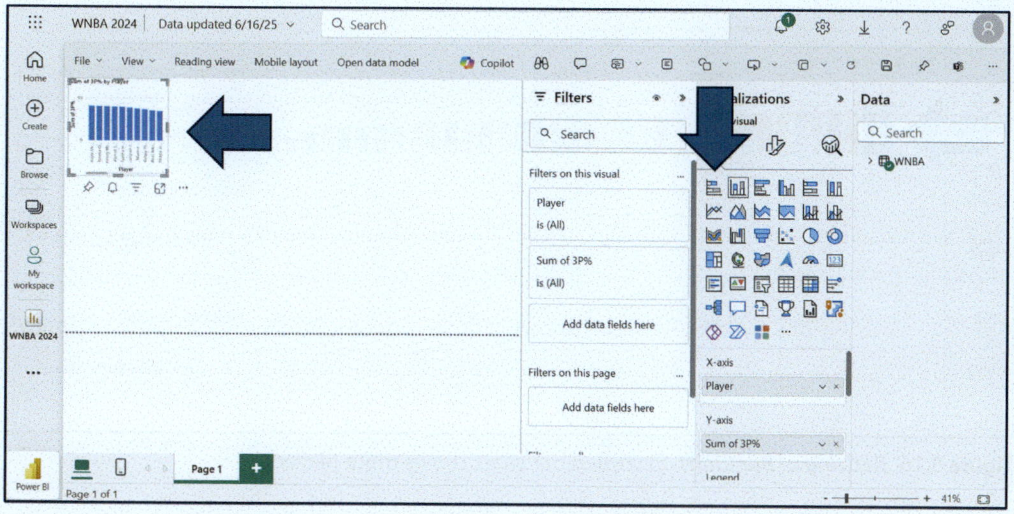

Figure 9.14 ***Chart Types Selection.*** Arrow pointing to Chart Types Selection.

14. Click the Stacked column chart icon. Drag Player onto Legend. The following appears (Fig. 9.15).

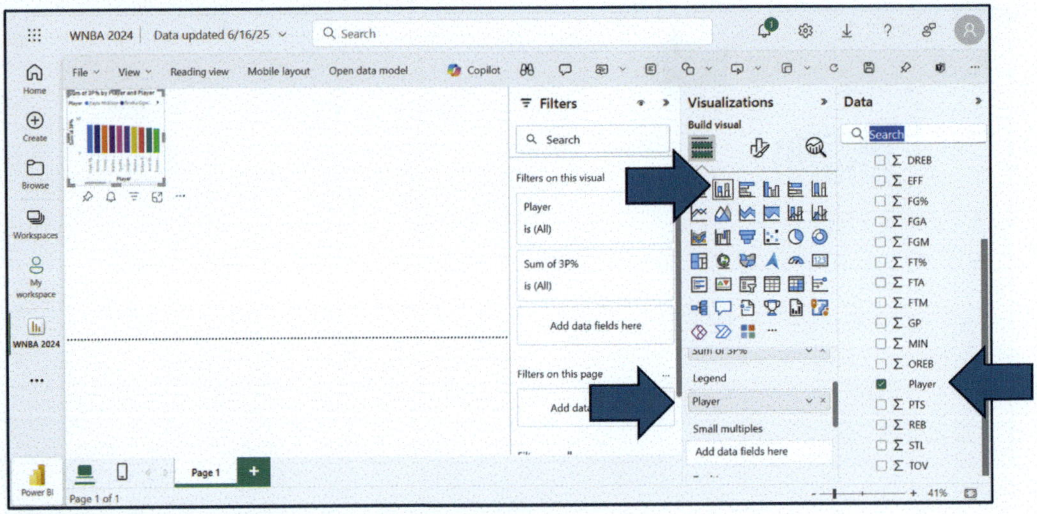

Figure 9.15 ***Stacked Column.*** Color-differentiated bars.

15. Resize the column chart to display additional players and their three-point shooting percentage (3P%) (Fig. 9.16).

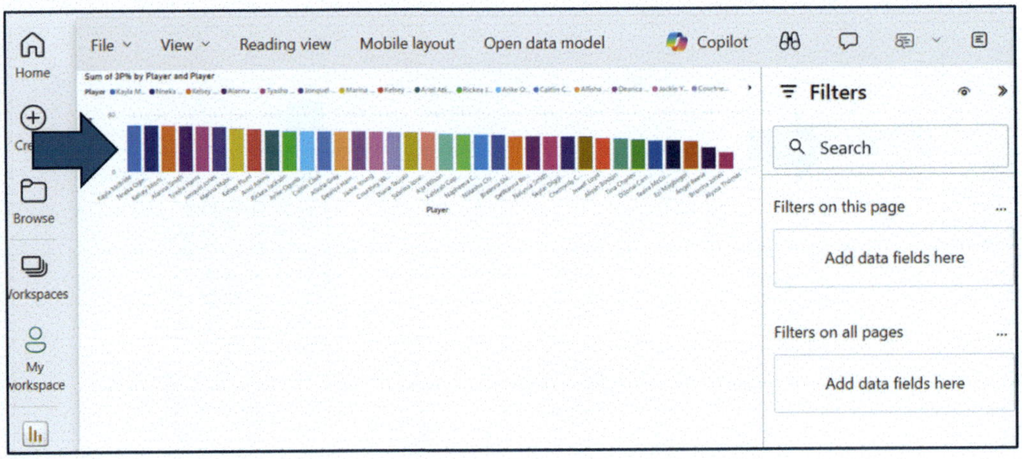

Figure 9.16 ***Resizing of Bar Chart.*** Extended bar chart shows more players.

16. To add a new graphic, first left click a blank space on the page (Fig. 9.17).

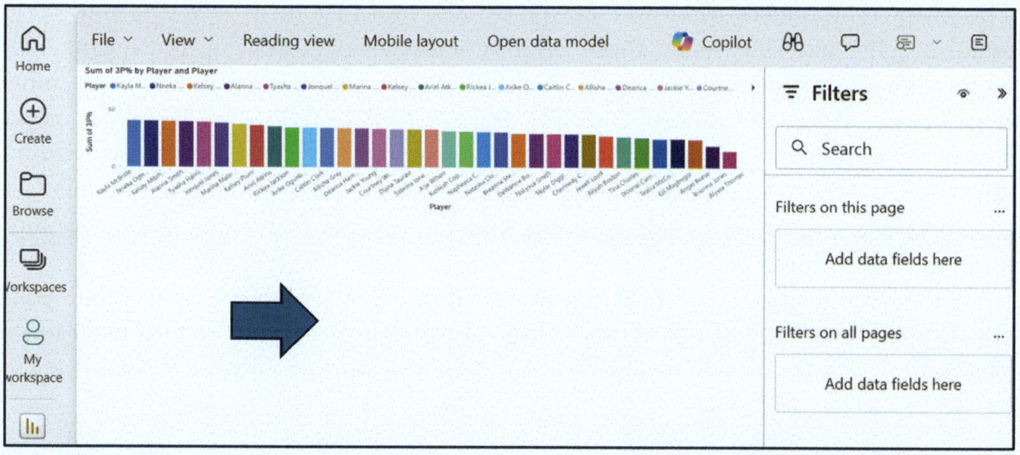

Figure 9.17 ***Blank Space Click.*** Arrow pointing to where you left click on a blank space.

17. Click the line chart icon under Visualizations. Under Data, select the player stat for field-goal percentage (FG%). A line chart will appear underneath the bar chart (Fig. 9.18).

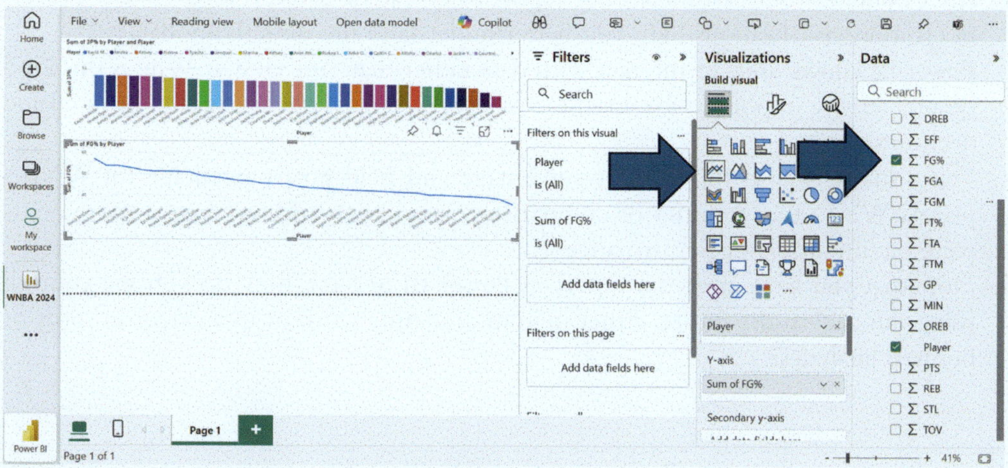

Figure 9.18 ***Line Chart.*** Line Chart below Bar Chart.

18. Save your work by clicking the icon for Save this report (Fig. 9.19).

Figure 9.19 ***Save Report.*** Highlights icon to "Save this Report."

SUMMARY

This chapter introduces Microsoft Power BI as a tool that turns raw data into clear, useful insights and helps share those insights across an organization. It explains how sport teams and organizations can use Power BI to track budgets, player statistics, injuries, fan attendance, and sponsorships. You learned how to upload an Excel file, select data fields, and build interactive visualizations directly in Power BI Service, Microsoft's browser-based platform. By completing these steps, you became familiar with the Power BI workspace, basic report-building tools, and the process of turning sport data into meaningful visual stories.

In the next chapter, you'll build on this foundation by learning how to refine your visuals, apply filters, adjust colors, and design pages that make your reports even more effective.

BEYOND THE BOX SCORE

1. John Tukey said that the greatest value of a picture is when it forces us to notice what we never expected to see. What surprised you when you visualized the WNBA data, and how might these insights influence decisions in sport?
2. After building your first interactive report in Power BI Service, how did the process of creating visualizations help you better understand the underlying data?
3. Power BI allows users to interact with data rather than just view it. Why is this kind of interactivity important for coaches, executives, or analysts in making sport-related decisions?
4. You visualized both three-point percentage and field-goal percentage for WNBA players. What other statistics or comparisons would you add to create a more complete picture of player performance?
5. Now that you have created a report in Power BI Service, how do you see yourself using this skill in the future, either in your academic work or in a career in sport or business analytics?

Mini Project

1. Write a 150–250-word reflection on the Power BI report created in this chapter.
2. Describe what the visuals show and what insights they reveal.
   Consider these guiding questions:
   - What patterns or trends stand out in the WNBA 2024 data?
   - Which players performed best in three-point and field-goal percentage?
   - How could coaches, analysts, or executives use this information to make decisions?
   - What was your experience using Power BI for the first time? What did you find easy or challenging about the process?

Focus on explaining what the visuals *mean* rather than how they were built. Your goal is to communicate the story behind the data clearly and professionally, just as a sport analyst would in an organization.

AI DISCLOSURE

During the preparation of this work, the author(s) used ChatGPT (OpenAI GPT-5) in order to provide editing and improve the readability of the document. After using this tool/service, the author(s) reviewed and edited the content as needed and take(s) full responsibility for the content of the publication.

REFERENCES

Knight, D., Robinson, B., Pearson, M., Moss, J., & Wallace, A., 2022. Microsoft Power BI quick start guide (3rd ed.). Packt Publishing.

Microsoft. 2024. Overview of Power BI capabilities. Available from: https://learn.microsoft.com/en-us/power-bi/fundamentals/power-bi-overview.

Panchal, K. 2024. Power BI vs Tableau: Which is better for your business? Available from: https://www.ifourtechnolab.com/blog/power-bi-versus-tableau.

Tukey, J. W., 1977. Exploratory data analysis. Addison-Wesley.

CHAPTER 10

Building Visuals in Power BI

Key Takeaways

- Customizing visuals in Power BI helps make data easier to read and understand.
- Features like filters and tooltips allow users to highlight key information and add helpful details without overcrowding the report.
- Good design in Power BI focuses attention on what matters most, so insights are clear and meaningful.

The ability to simplify means to eliminate the unnecessary, so that the necessary can speak.
– Hans Hofmann (Hofmann, 1967)

In this chapter, you'll begin customizing data visualizations using Power BI Service. You'll learn how to build and personalize charts by filtering data, adjusting labels, changing colors, and adding design elements like background images, titles, and borders. These tools will help you turn raw data into a clear, engaging report that highlights key insights.

DESIGN ELEMENTS IN POWER BI SERVICE

1. Open Power BI Service at app.powerbi.com. Open the WNBA 2024 report that was created previously.

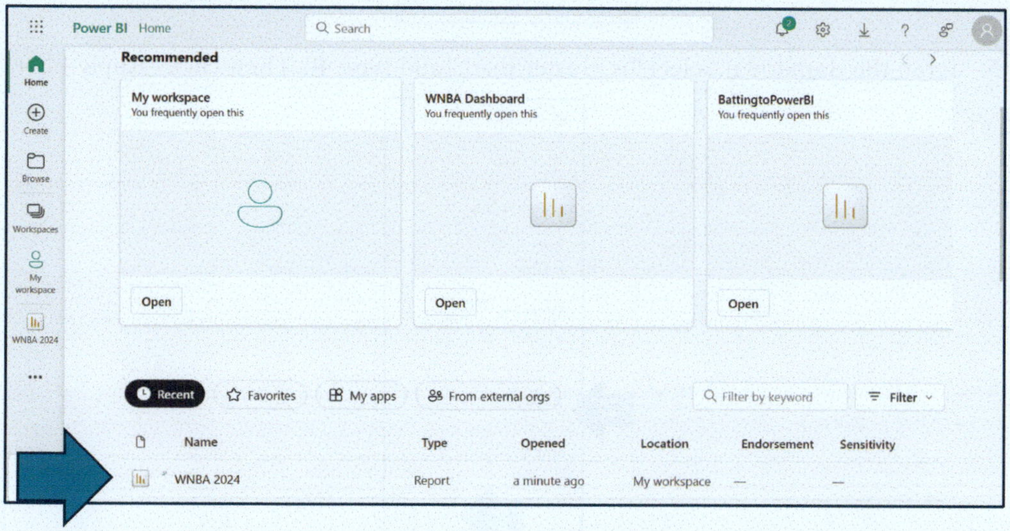

An Introduction to Applied Sport Analytics **https://doi.org/10.1016/B978-0-443-49068-2.00010-0**
© 2027 Elsevier Inc. All rights are reserved, including those for text and data mining, AI training, and similar technologies.

2. You may need to click the Edit icon to access the report.

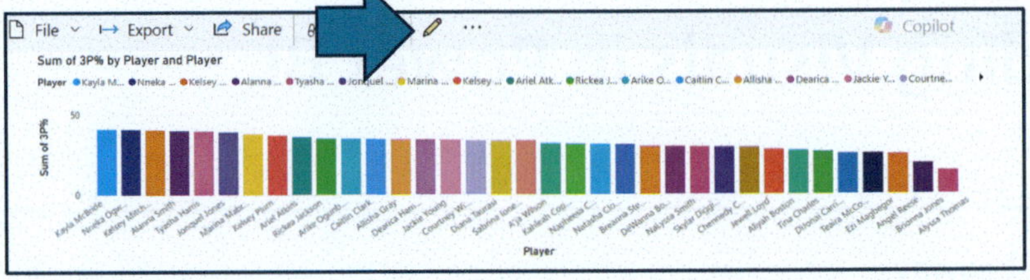

3. Left-click the column chart titled "Sum of 3 P% by Paper and Player."You'll notice the appearance of a bounding box with handles around the chart.

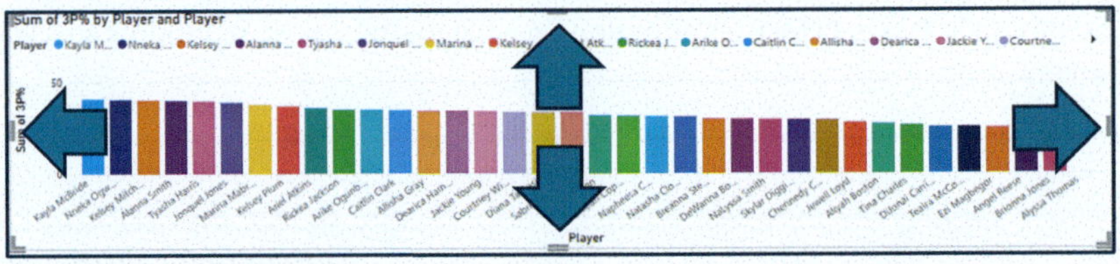

4. Let's reduce the number of players shown in the chart. Under Filters, expand the filter for Sum of 3 P% by clicking the dropdown arrow.

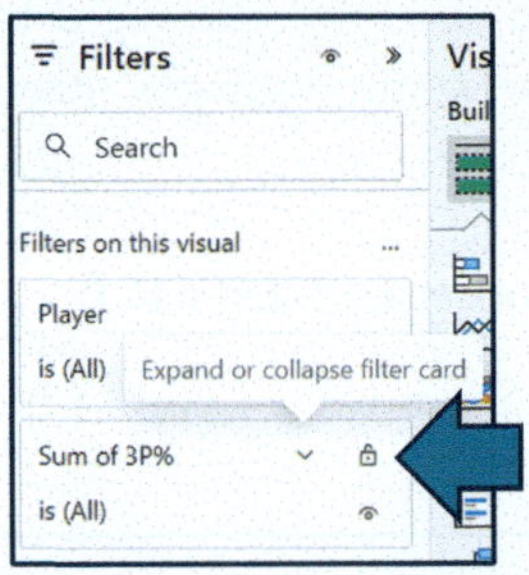

5. From the dropdown, select "is greater than," and type 40. Then click "Apply filter."

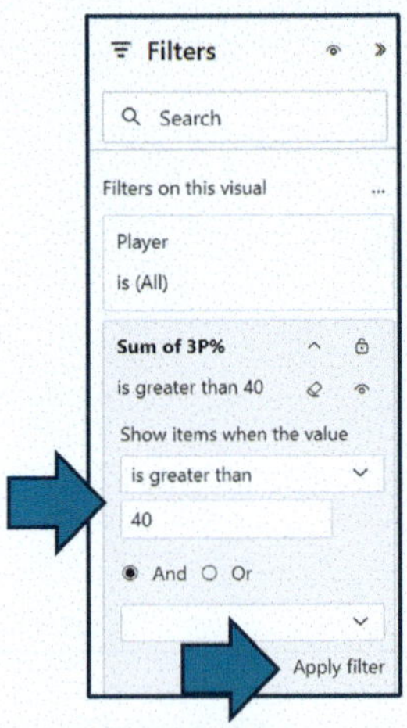

6. The column chart now shows just the top three players in the WNBA. These are the three players who shot three-pointers at a rate better than 40% in 2024.

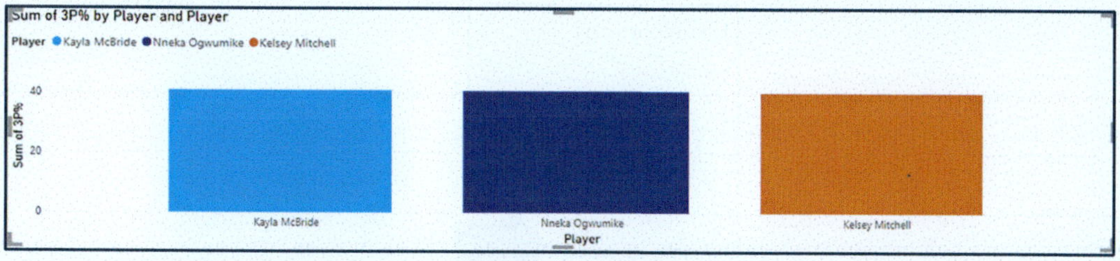

7. Let's change the title of the chart. Under Visualizations, select the icon for "Format visual," then click General. Click the dropdown for Title. In the Text cell, type "Top 3 in 3 P%." Change the font to Arial, 18 point, and bold.

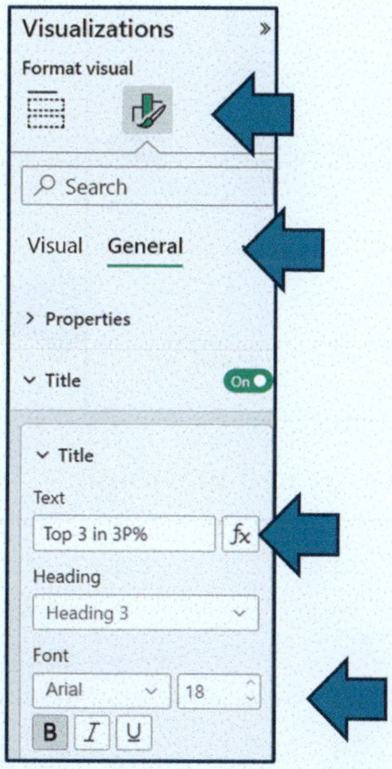

8. Click Visual. Turn the Legend off. The legend disappears from our chart.

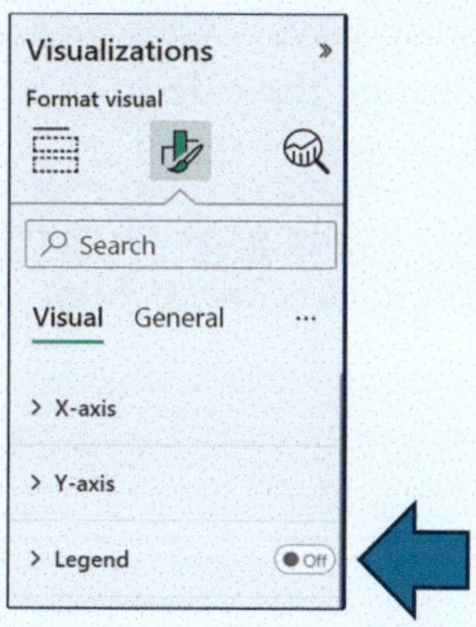

9. Go to General > Effects. Change the background color of the chart. Give the background a 50% transparency.

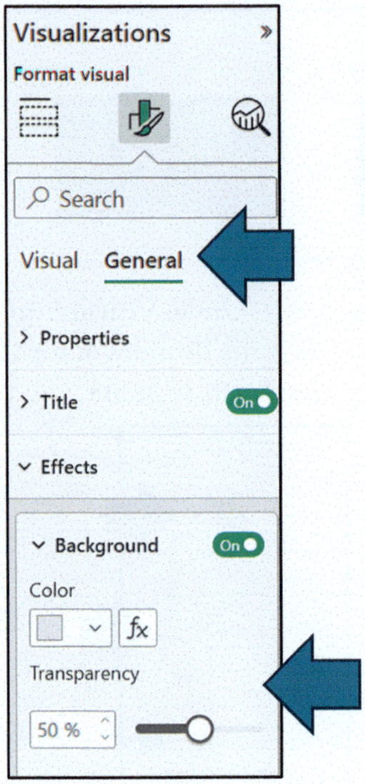

10. Select the line chart that we created previously.

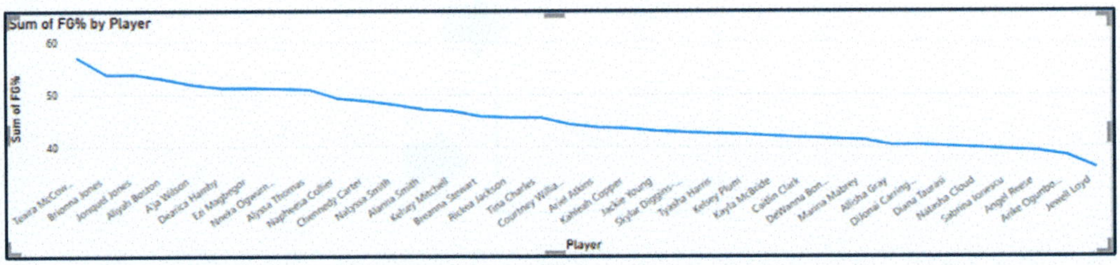

11. Adjust the filter to only display players who shot greater than 50% on field goals (see step 5 for assistance). Change the title of the chart to "Top Players for FG%" (see step 7). Adjust the background of the chart (see step 9).

12. Let's add a background image to the page. Drag the two visualizations to the bottom of the page. Deselect the charts we've been working on by clicking a blank area on the page.

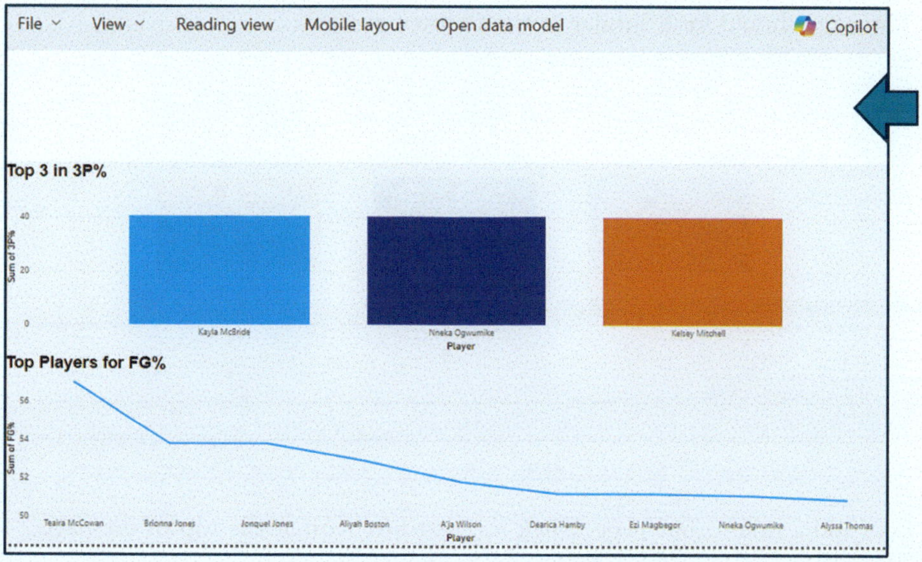

13. Download the file "Basketball_Image.jpg" from the book's website. In Power BI Service, select the icon for "Format page," select "Canvas background," and then under Image navigate to select "Basketball_Image.jpg." For Image fit, select Fit. Set Transparency to 75%.

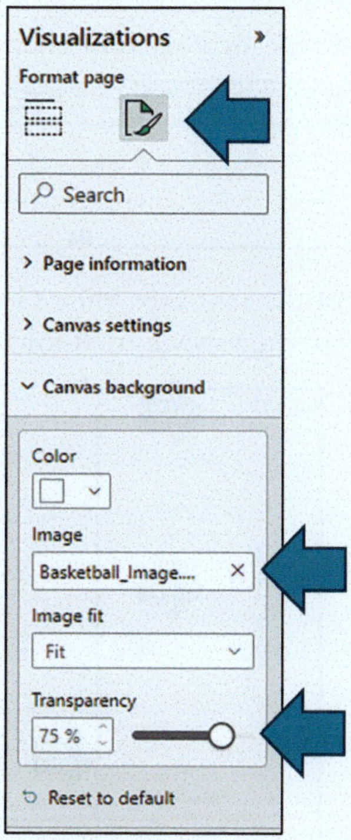

14. The page should look similar to the following.

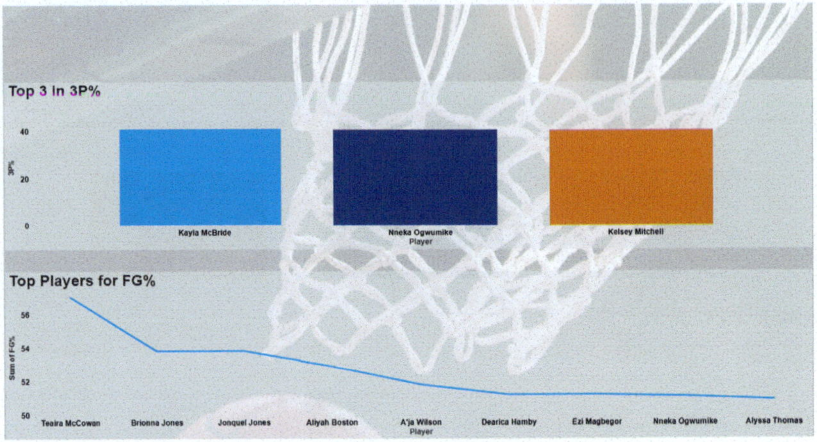

15. Let's add a title to our page. Click the text box icon at the top of the page.

16. Give the page a title such as "Top WNBA Players in 2024." Select 36-point Arial font in bold. **Note**: You will need to select or highlight the text "Top WNBA Players in 2024" in the text box for your font changes to take effect.

17. To adjust the background of the text box, click General and select the dropdown for Effects. Apply the same settings as we used for the charts.

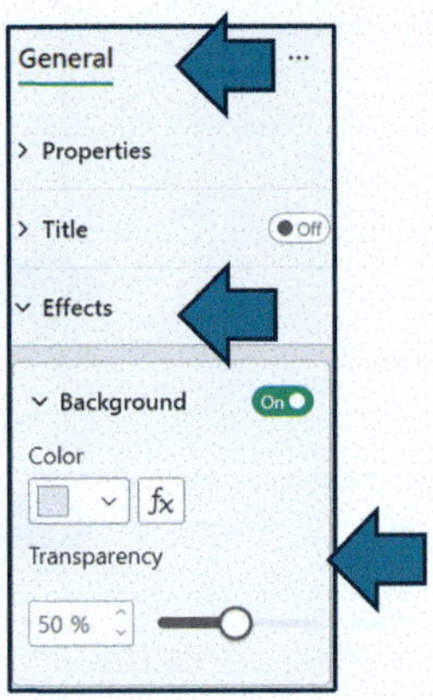

18. Let's give the text box a border. With the text box still selected, under General, turn on Visual border. Give the border 10-point rounded corners with a 5-point width.

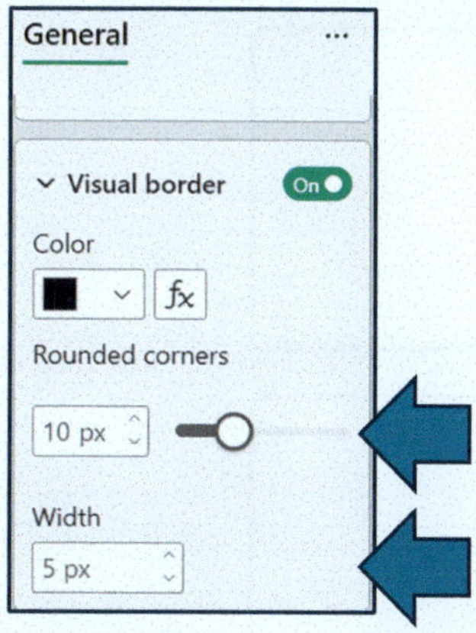

19. For consistency, give the charts these same settings for the borders. First select a chart, then under Visualizations select the icon for "Format visual." Select General and give the Visual border the same settings that we applied to the text box.

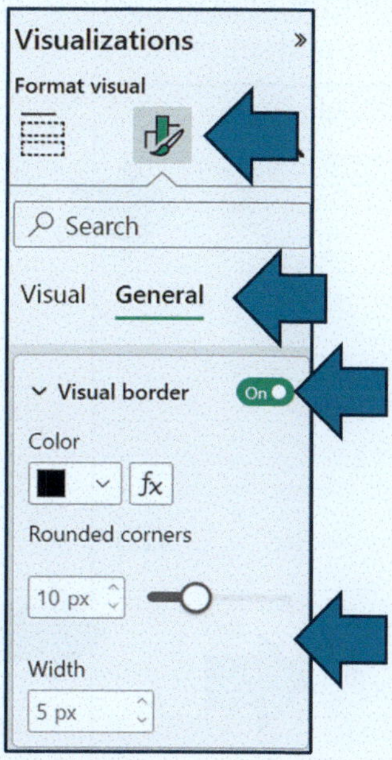

20. Apply these same settings to the second chart.

21. Reposition the title and charts for visual appeal.

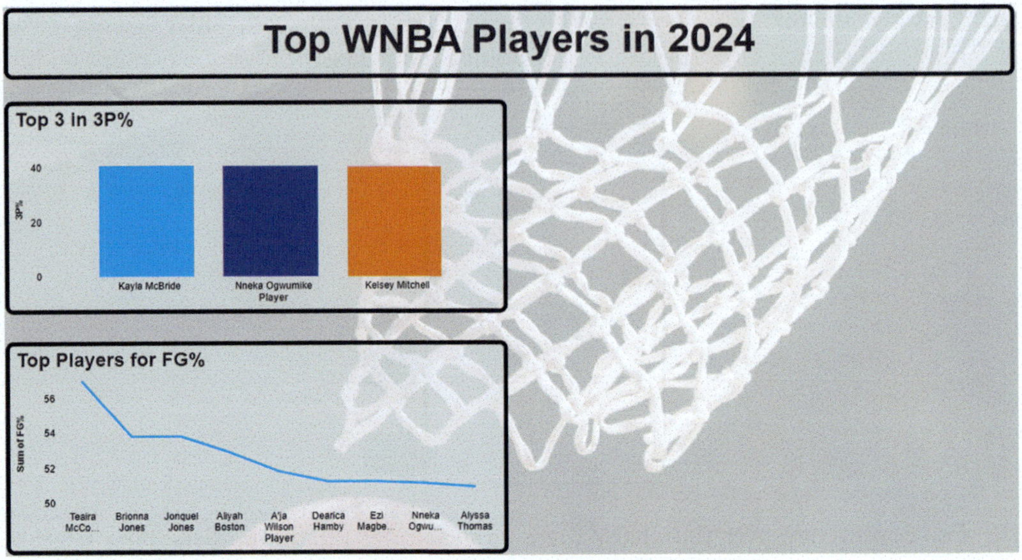

22. Let's adjust the fonts for our x-axis and y-axis. Select the bar chart, then under Visualizations, click the icon for "Format visual," click the dropdown for x-axis, and make the font Arial, 14 point, in bold.

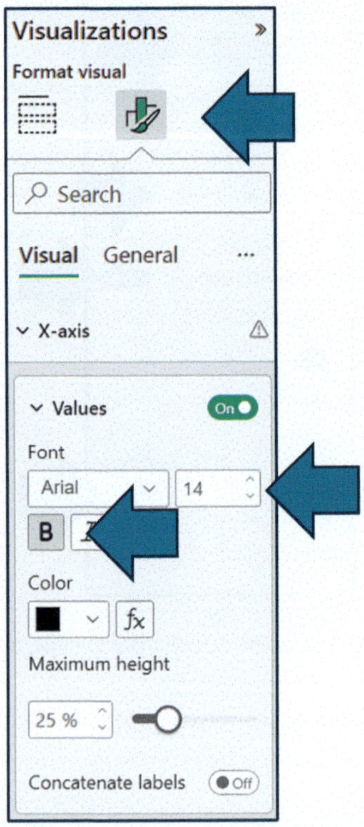

23. Under Title, apply the same 14-point Arial font in bold.

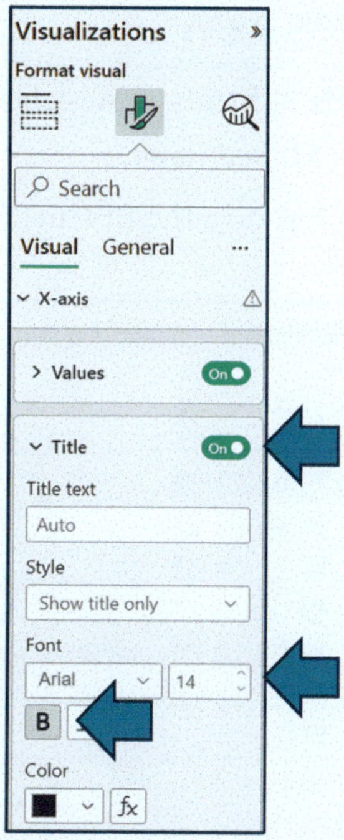

24. Apply these same settings to the y-axis.
25. Apply these same settings to the line chart.

You now have the beginnings of your first report using Power BI. What changes would you make to this page to improve its visual appeal (backgrounds, colors, fonts, transparency, spacing, etc.)? What additional charts might you add to make it more informative? Think about the key questions your report should answer and which visualizations best highlight those insights. Consider adding comparisons, trends over time, or team-based summaries to deepen your analysis.

TOOLTIPS

You'll notice that when you mouse over the bar charts, a tooltip will appear that displays information about the individual player. In this instance, we can see that Kayla McBride made 40.70% of her three-point attempts in 2024. As you'll find with most features in Power BI, tooltips can be customized to provide the information that you want to display.

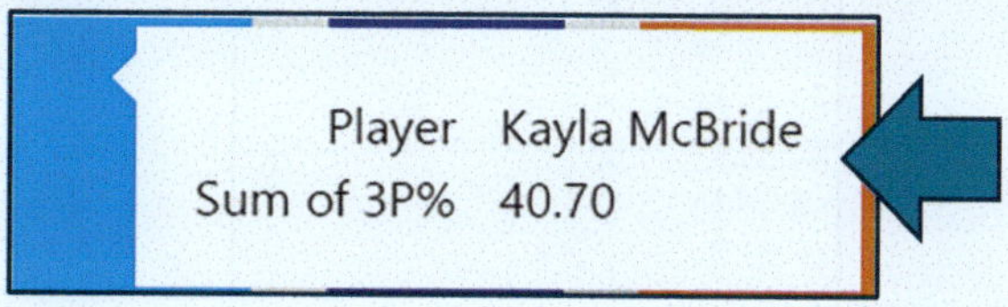

Let's say we also want to provide each player's efficiency (EFF) score on the tooltip. EFF is a linear statistic that measures a player's overall production. EFF provides a single number summarizing a player's positive contributions and negative contributions. Here is the formula for EFF (NBAstuffer, 2025):

$$\text{Points} + \text{Rebounds} + \text{Steals} + \text{Assists} + \text{BlockedShots} - \text{Turnovers} - \text{MissedShots}$$

1. To edit the content of the tooltip, drag the EFF field to "Add data fields here" under Visualizations > Tooltips.

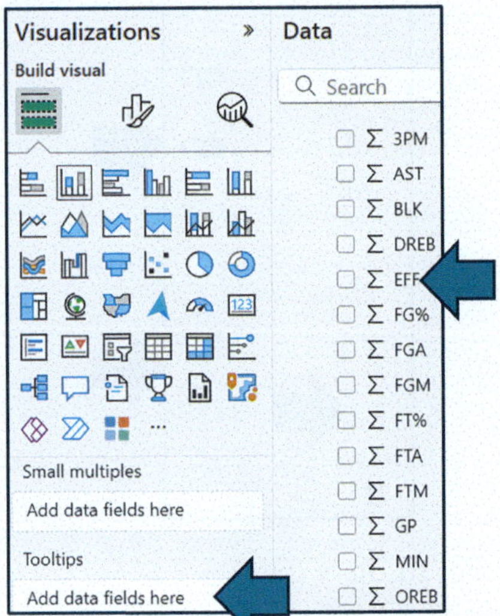

2. Now, when you mouse over Kayla McBride, you'll see the addition of "Sum of EFF" to the tooltip.

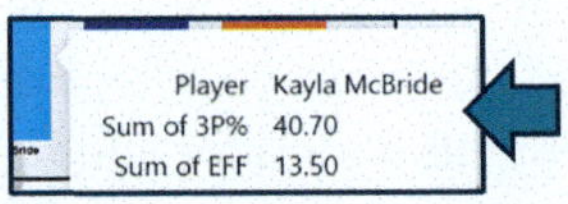

3. To change the label from "Sum of EFF" to "EFF," under Tooltips, click the drop-down arrow next to "Sum of EFF," and select "Rename for this visual."

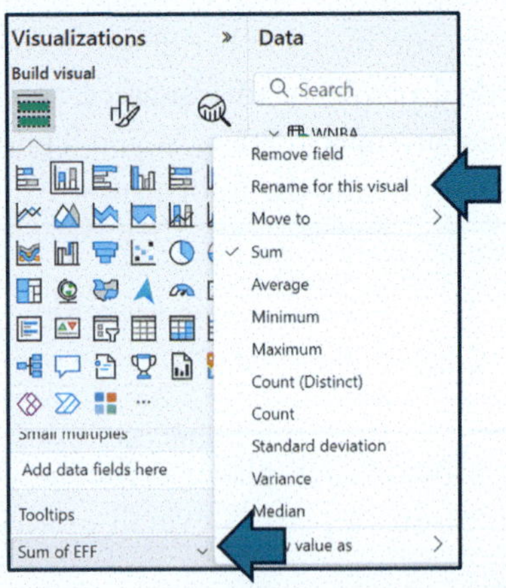

4. Type "EFF" in the cell and press [Enter]. The tooltip has been modified.

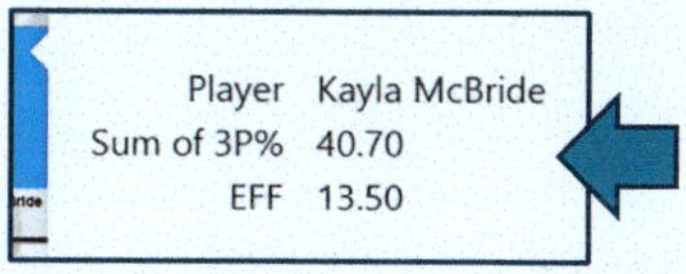

5. To change the appearance of the tooltip, click the "Format visual" icon, click General, and then select the dropdown for Tooltips. For example, under Text, you can change the font color to white, and under Background, you can change the background of the tooltip from white to black.

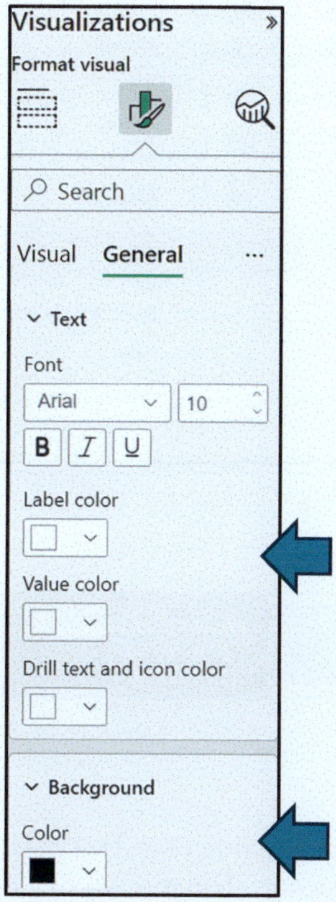

By performing a mouse over of the top three shooters in three-point percentage, you will see that Nneka Ogwumike has the highest EFF score. This indicates that, among these three players, Ogwumike contributes the most overall across multiple statistical categories. This reinforces what we know of this nine-time WNBA All-Star.

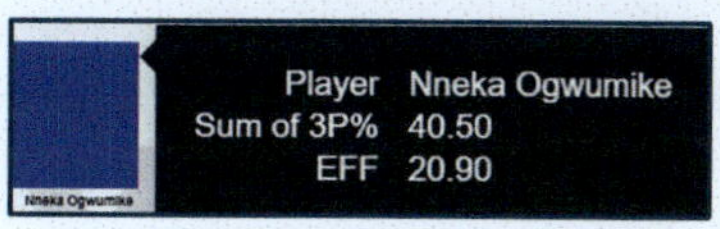

6. To change "Sum of 3P%" on the tooltip to just "3P%," you can follow the same process by changing the label under y-axis.

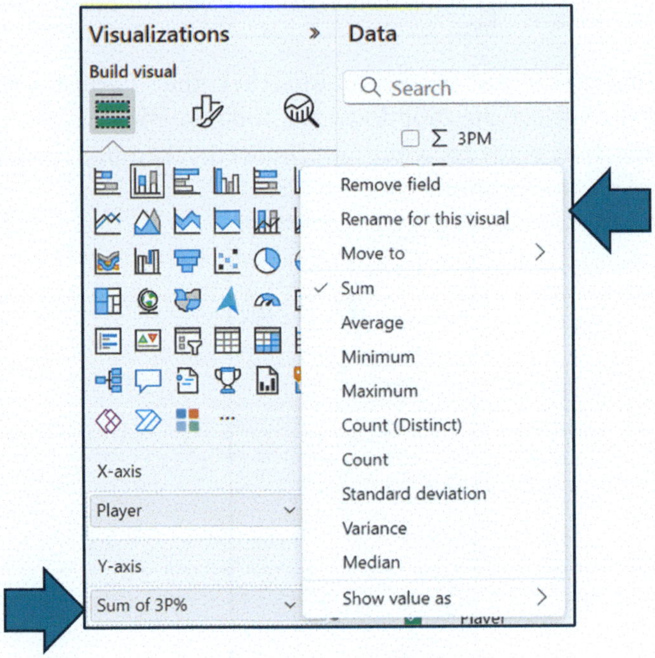

The tooltip is now easier to read.

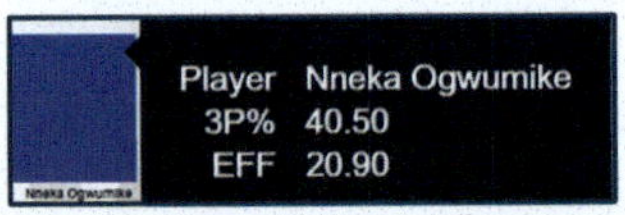

SUMMARY

In this chapter, you explored design choices in Power BI. By customizing visualizations with filters, labels, colors, and layout elements, you can draw attention to key insights and improve clarity. This chapter also introduces tooltips, a powerful way to enhance Power BI reports by displaying additional information when users hover over visuals. Together, these techniques help focus attention on what matters most and allow users to gain insights from data.

BEYOND THE BOX SCORE

1. Tooltips allow viewers to explore additional data without overwhelming the main report. In what ways does this support Hans Hofmann's idea that "the ability to

simplify means to eliminate the unnecessary so that the necessary may speak"? How does this feature improve the clarity and usefulness of a visualization?

2. You added a player's EFF score to the tooltip. Why is it helpful to include secondary statistics like EFF, and how can this enhance the viewer's understanding of player performance?
3. Formatting the tooltip made it easier to read. What design principles apply to tooltip customization, and why do font color, background color, and naming conventions matter in analytics presentations?
4. You filtered and formatted charts to highlight top-performing WNBA players. Why is it important to reduce visual clutter and focus attention in a report, and how can formatting choices support that goal?
5. Design elements like titles, fonts, colors, and borders can change how a visualization is perceived. Which design choices made your report easier to interpret, and which ones do you think were more about style than substance?

Mini Project

1. Create a visual in your report using a statistic not previously featured (e.g., assists, rebounds, or turnovers).
2. Customize your visual by adjusting the title, colors, labels, and background to make it clear and professional.
3. Write a 150–250-word explanation describing your visual and what it reveals. Consider these guiding questions:
   - What data did you choose to highlight, and why?
   - What insights or trends does your visual show?
   - How did your design choices (such as color, layout, or filters) help communicate the message?
   - How could coaches, analysts, or executives use this information in decision-making?

AI DISCLOSURE

During the preparation of this work, the author(s) used ChatGPT (OpenAI GPT-5) in order to provide editing and improve the readability of the document. After using this tool/service, the author(s) reviewed and edited the content as needed and take(s) full responsibility for the content of the publication.

REFERENCES

Hofmann, H., 1967. Search for the Real and Other Essays. Weeks, S.T. (Ed.). MIT Press. (Original work published 1948).

NBAstuffer, n.d. NBA efficiency. *Analytics 101*. Available from: https://www.nbastuffer.com/analytics101/nba-efficiency/ (accessed 28. 06.25).

CHAPTER 11

Analyzing Data With Power BI

Key Takeaways
- Slicers make reports interactive and provide focus on key data.
- Power BI features such as forecasting tools show possible future trends and help predict outcomes.
- Sharing reports lets others explore and use your insights.

Data visualization is the language of decision making. Good charts effectively convey information. Great charts enable, inform, and improve decision making.

– Dante Vitagliano (Alex, 2024)

In this chapter, you'll learn how to use slicers to make your Power BI report more interactive and user-friendly. Slicers allow users to filter data directly within a report or dashboard by selecting specific values such as players, teams, or categories. This makes it easy for decision-makers to focus on the information that matters most to them.

1. Let's walk through how a slicer works in Power BI. First, let's create a table by clicking the Table icon.

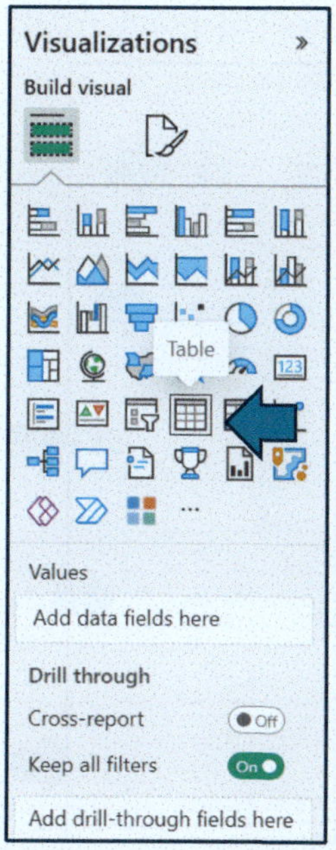

An Introduction to Applied Sport Analytics https://doi.org/10.1016/B978-0-443-49068-2.00011-2
© 2027 Elsevier Inc. All rights are reserved, including those for text and data mining, AI training, and similar technologies.

2. Populate the table with all players (Player) in the WNBA in 2024 and their defensive statistics for blocks (BLK), defensive rebounds (DREB), and steals (STL). Put a check box next to each of these variables.

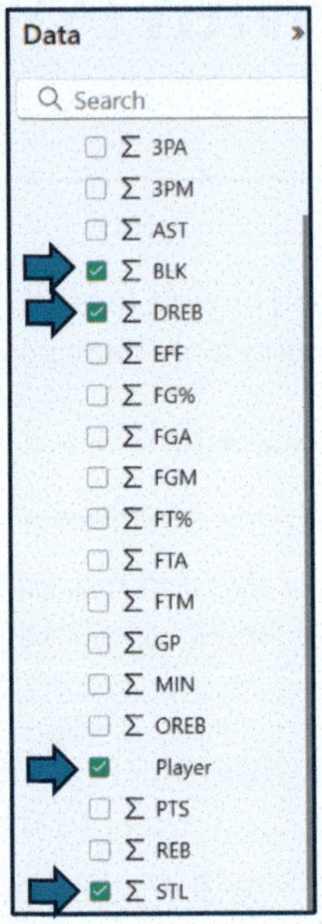

3. Click the table; then under Visualizations, click the "Format visual" icon, click General, and turn on the Title function. Type "Defensive Stats" under Text, and give the title the same characteristics (18-point, Arial font in bold) as our other titles.

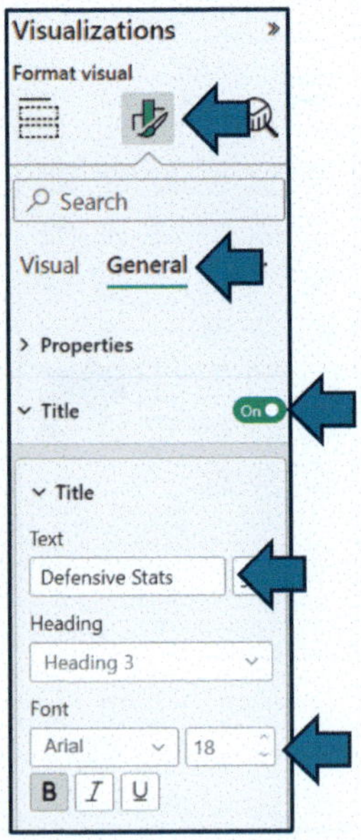

4. Under Background, give this table the same characteristics as our other charts.

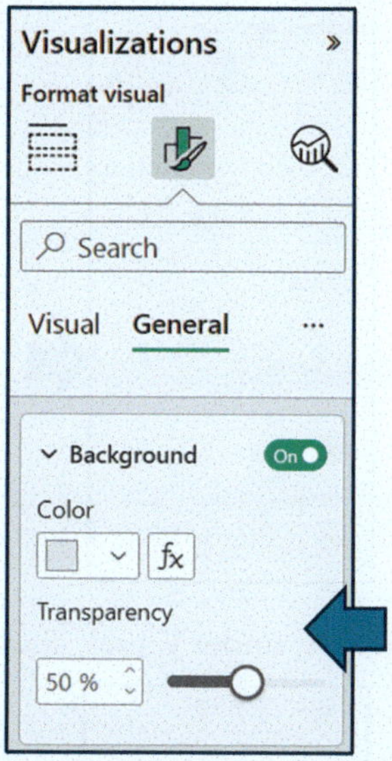

5. For Visual border, give the chart the same characteristics as our other charts.

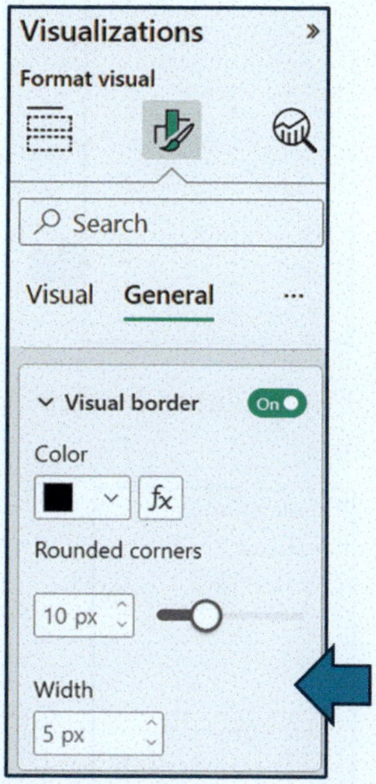

6. Under Visual, turn off the option for Totals. This removes totals for blocks, defensive rebounds, and steals from the table.

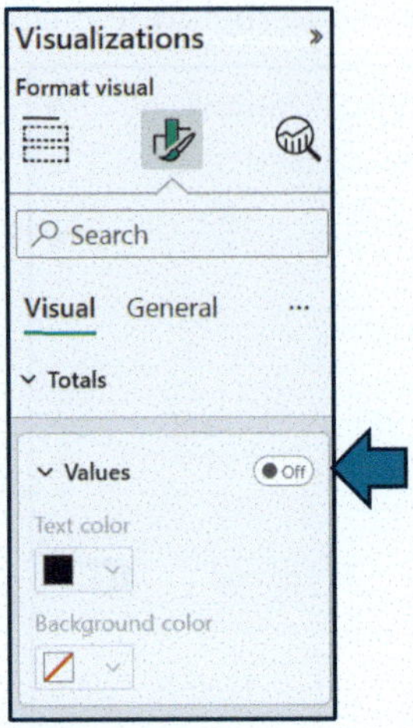

7. To make the text more readable, under Values, change font to Arial and 14 point.

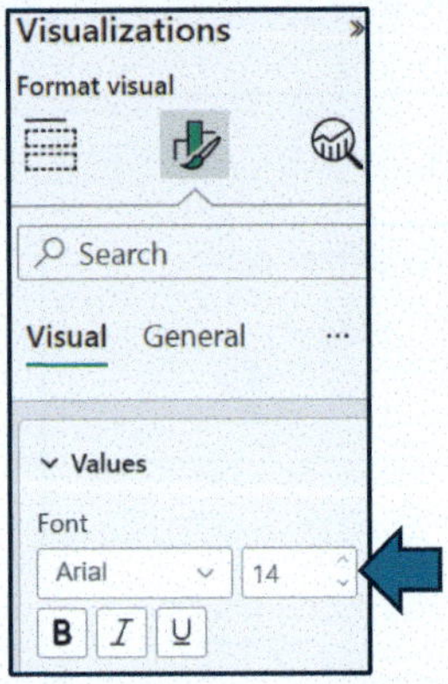

8. Let's also change the font for our column headers. Under Column headers, change the font to Arial and 14 point.

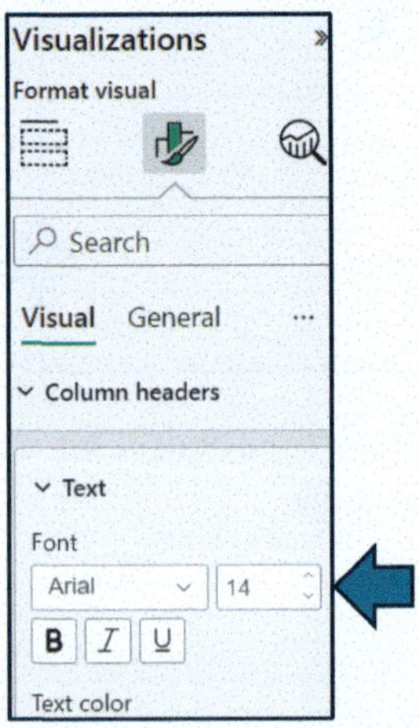

9. We should change the text for our column headers as well. Click the icon for "Add data to your visual." Change "Sum of BLK" to "BLK," "Sum of DREB to DREB," and "Sum of STL" to "STL." To do this, click the down arrow next to each option, and then choose "Rename for this visual."

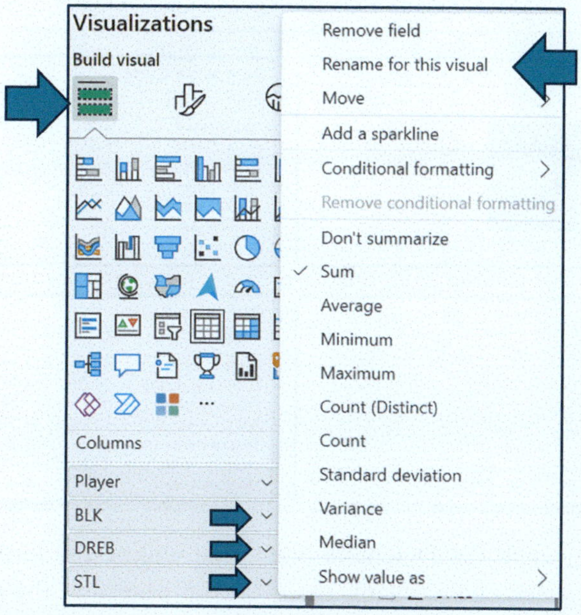

1. Resize your table so that your page looks like this.

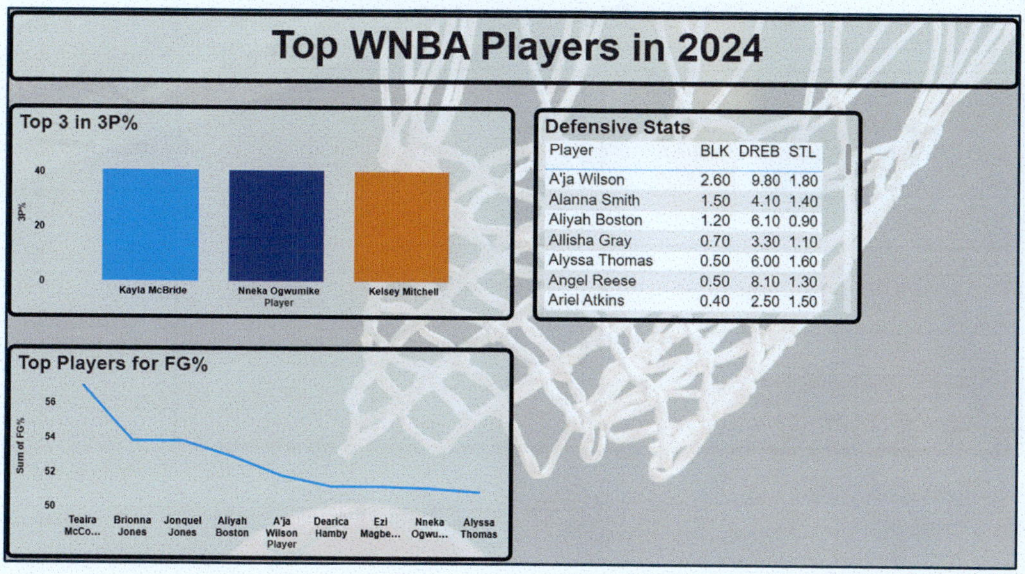

2. Let's add a table called Offensive Stats. Click a blank area on the page and add another table. Follow the steps we used to create the Defensive Stats table to make a new table for Offensive Stats. The page should look like this.

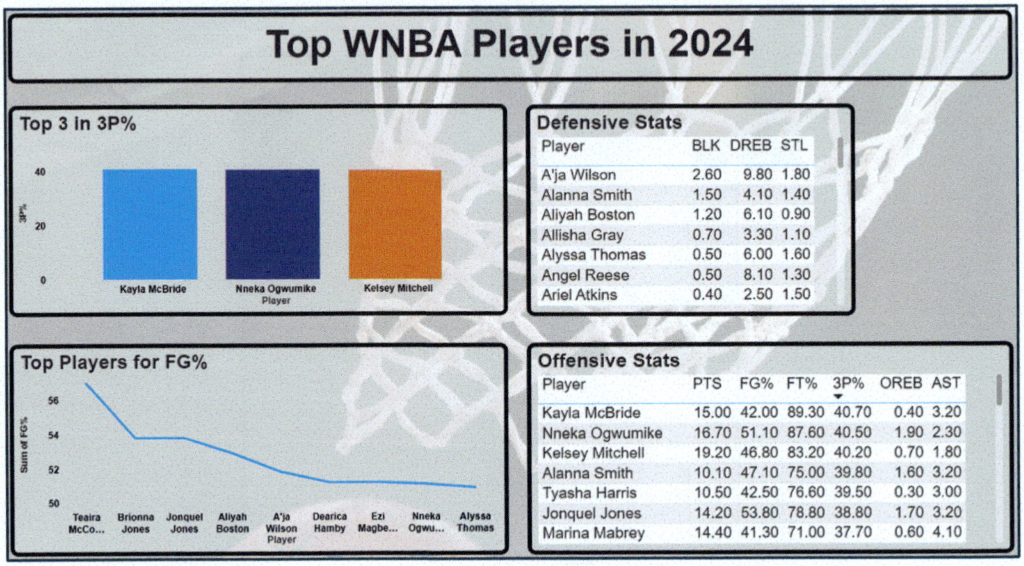

3. The space next to the Defensive Stats table will be the location for our slicer. To insert a slicer, click a blank space on the page, and then choose the Slicer icon.

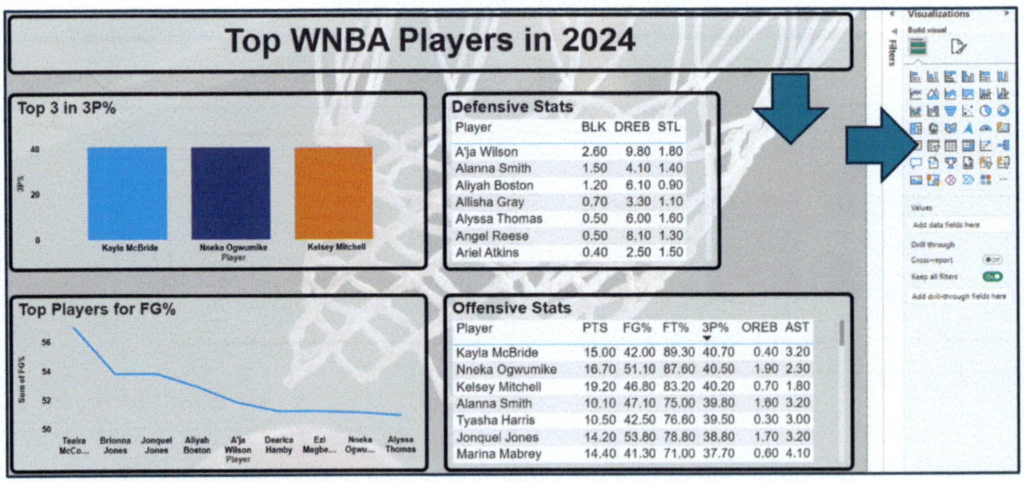

10. Under Data, put a check mark next to the variable Player, and reposition the slicer next to Defensive Stats.

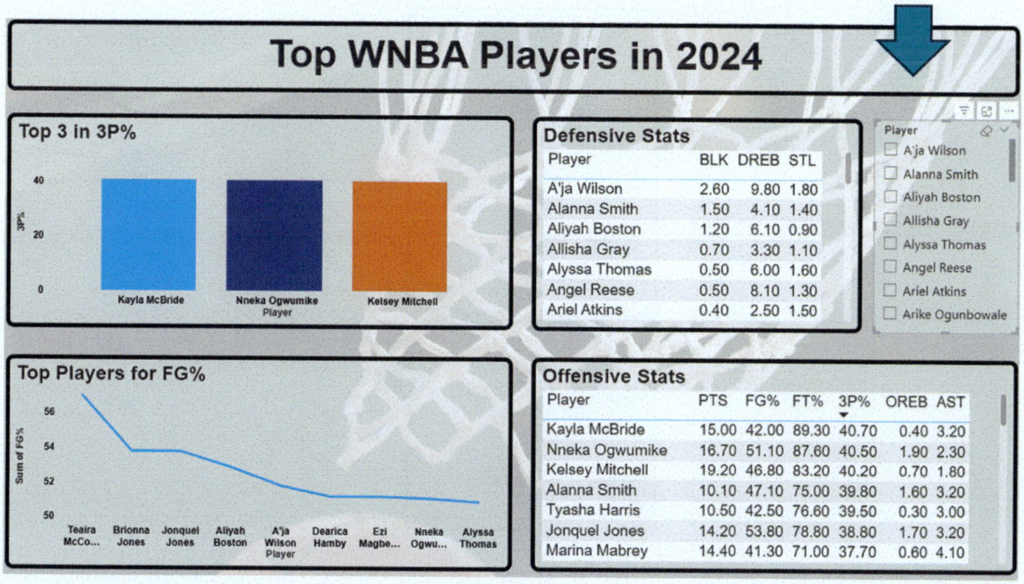

11. For consistency, click the "Format visual" icon, click General, and change the Visual border option to 10-point rounded corners with 5-point width.

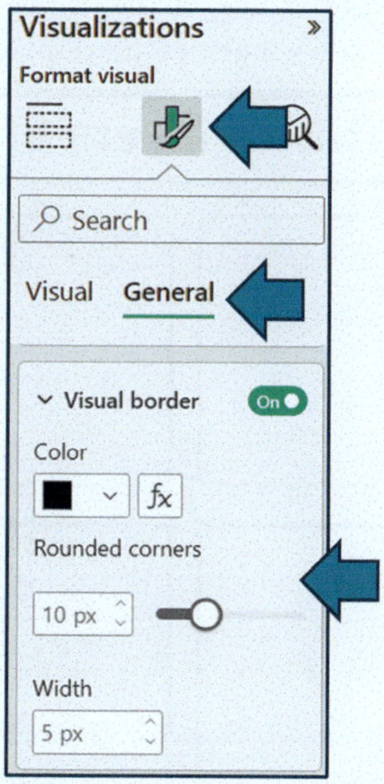

12. Under Background, adjust the background of your slicer to match the settings used for the rest of the page.

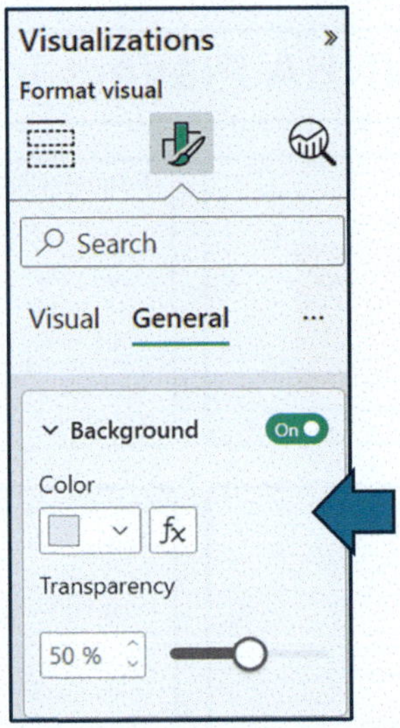

13. In the slicer, put a check mark next to Aliyah Boston. You will notice that since she's not in the top three as a three-point shooter, the bar chart disappears. She is, however, one of the top field-goal shooters by percentage as represented by the blue data point in what was the line chart. In the Defensive Stats and Offensive Stats tables, only Aliyah Boston's statistics are displayed.

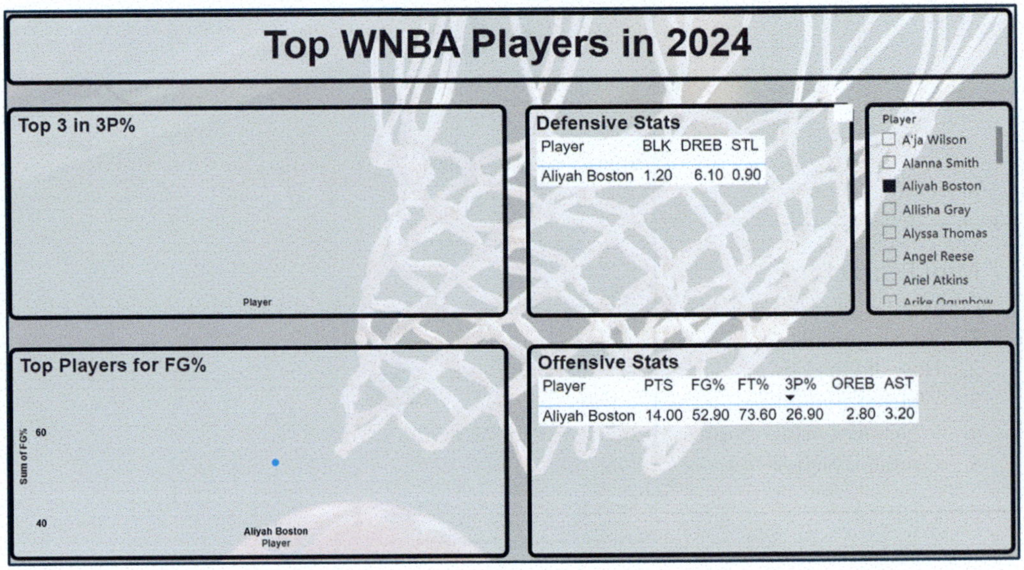

14. To make all data reappear, uncheck the box in the slicer next to Aliyah Boston's name.

POWER BI ANALYTICS

Power BI Service includes an analytics tool that can provide insights into charts and visuals. The Analytics pane only appears when a visual is selected on the Power BI page. When selecting this option, a variety of reference lines can be added to charts, including:

- X-axis constant line
- Y-axis constant line
- Minimum line
- Maximum line
- Average line
- Median line
- Percentile line

Power BI's analytics functionality also includes forecasting for time-series data. This feature helps project future values based on historical trends, which is especially useful when using line charts. Forecast settings let you adjust how far into the future you'd like to project and set the level of confidence in the prediction.

FORECASTING IN POWER BI

Let's create a second page using WNBA salary cap data to demonstrate how to use forecasting in Power BI (Spotrac, n.d.). Download the file "WNBA Salary Cap.xlsx" from the book's website. Create a new page by importing this file. Refer to the steps shown in Chapter 9 for assistance.

1. Under Visualizations, select line chart.

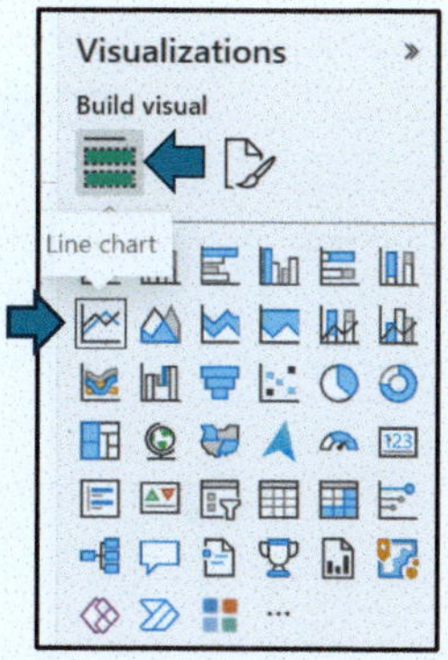

2. Under Data, put check marks next to Cap Max and Year. A line chart will appear.

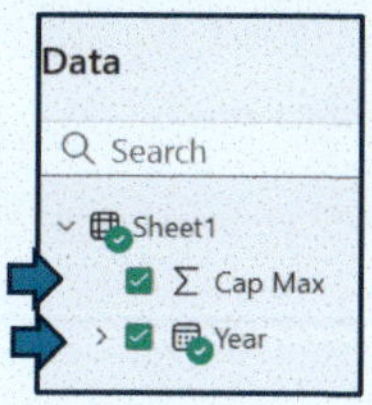

3. Under Visualizations, click the magnifying glass icon to access Analytics functionality within Power BI.

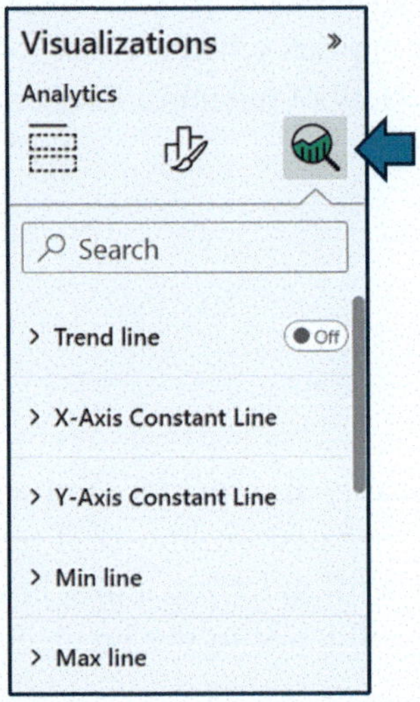

4. Turn on Forecast and expand this section by clicking the dropdown arrow. Change the forecast length to 5 years and click [Apply].

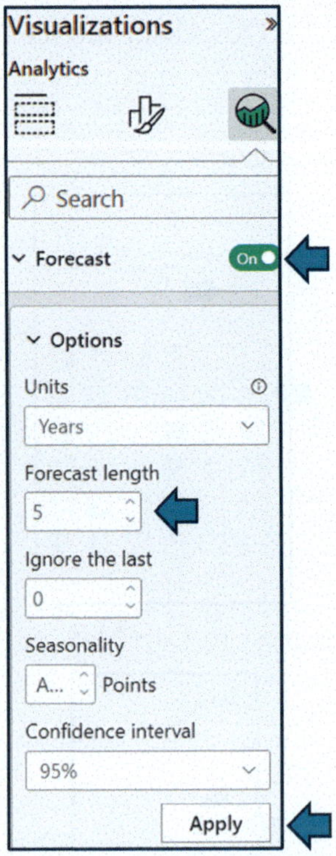

5. A 5-year forecast is added to the line chart. The gray band represents the confidence interval, which in this case was set at 95%. This means that if the forecast were repeated, the interval would contain the actual value in 95% of the cases (Visualitics Team, 2023).

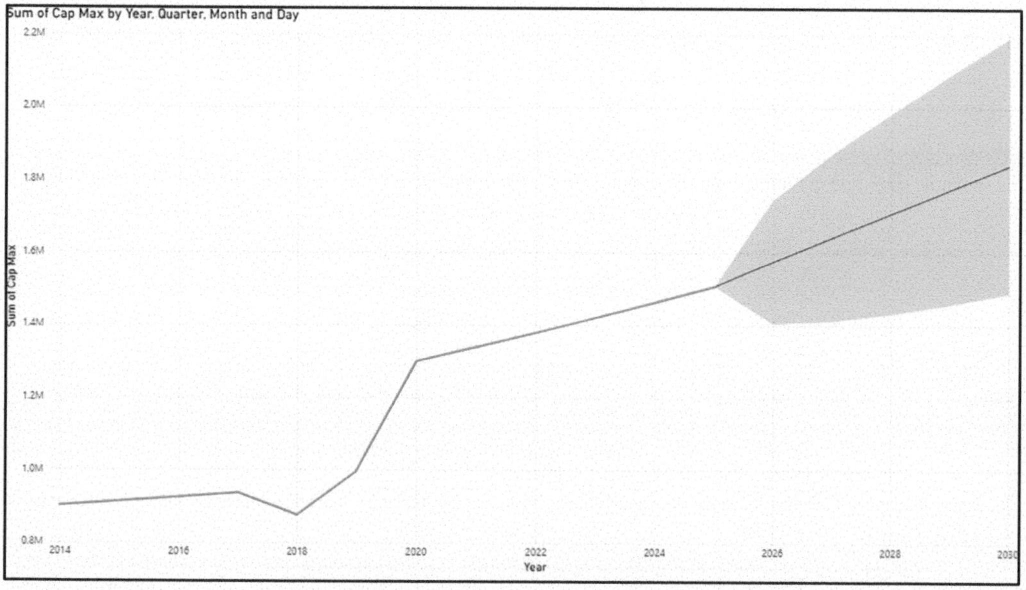

By using the mouse over functionality with the chart, you will see that by 2030, the salary cap for the WNBA is forecast to be $1.84 million per team, with a projected range of $1.49 million to $2.19 million.

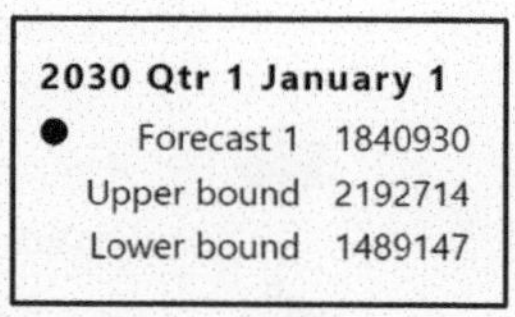

6. Save your work by clicking the Save icon.

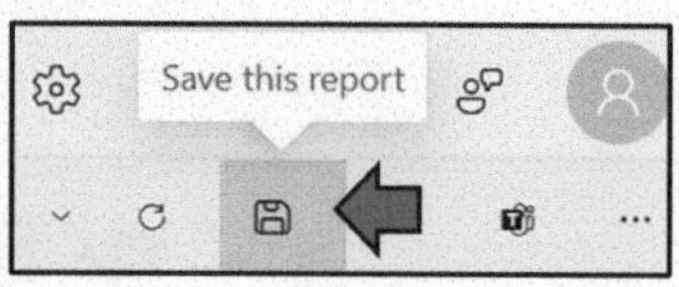

SHARING YOUR WORK

Now it's time to share your work, whether with the world or just a colleague down the hall. One of the powerful features of Power BI Service is the ability to publish and distribute your pages, reports, and dashboards for others to view and interact with. Power BI makes it easy to communicate your insights.

You'll want to pin your work to a dashboard so it can be shared. Begin by opening a page you created in Power BI Service. In the following case, we'll be using the forecast created previously that was saved as WNBA Forecast.

1. Click the icon to pin the page to a dashboard.

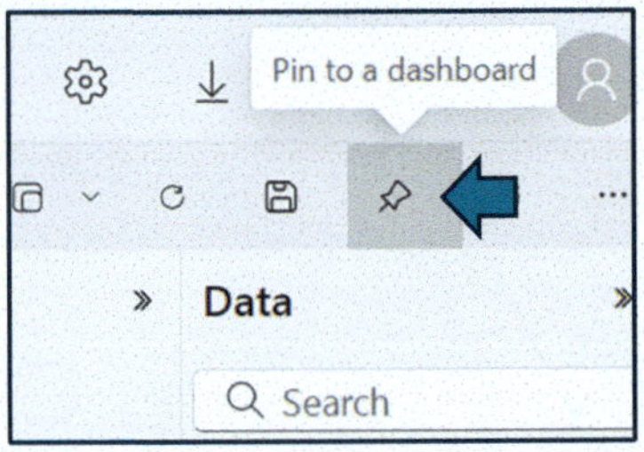

2. Choose the "New dashboard" radio button, give the dashboard a name, and click [Pin Live].

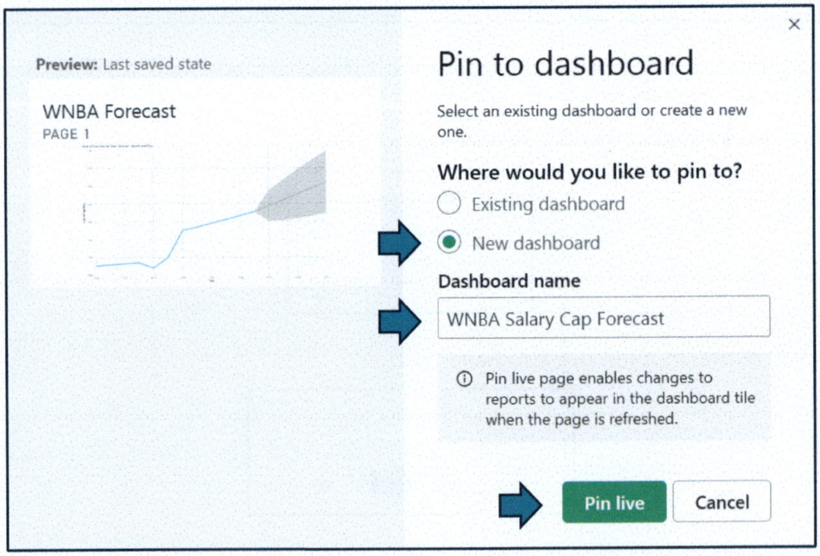

3. Click "Go to dashboard."

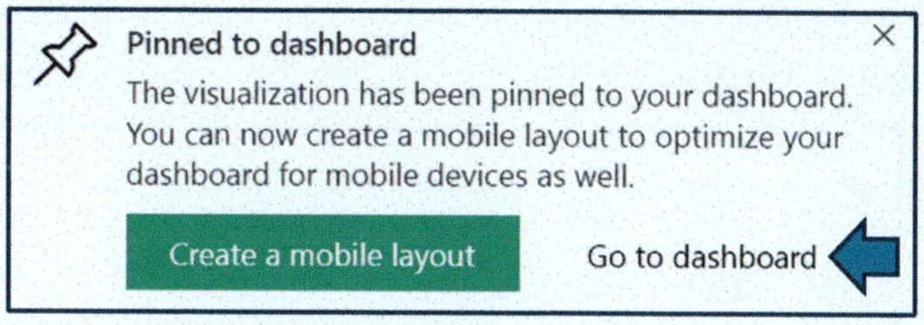

4. Click the full-screen mode icon to display your visual without distractions like menus and navigation panes. Press the escape [Esc] key to exit full-screen mode.

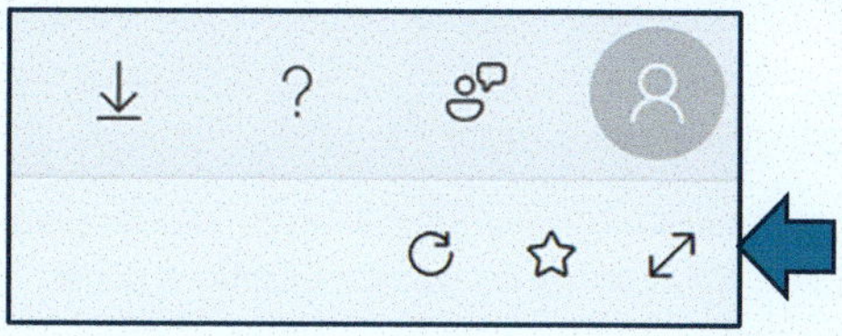

5. In order to share the visual with someone, click the Share icon.

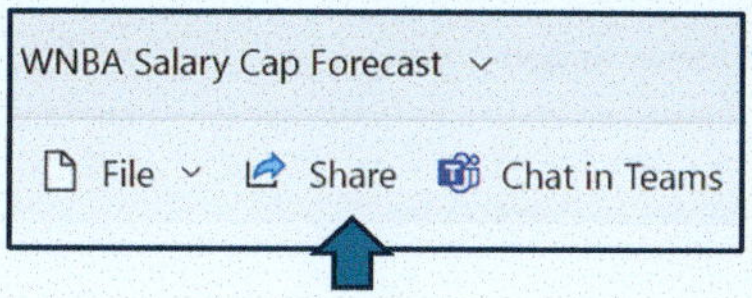

6. Type the individual's email address, select the check boxes you prefer, and click [Grant access].

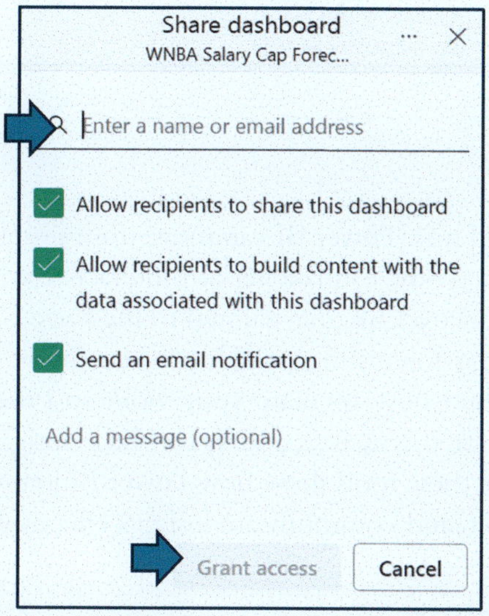

Note: To access the visual, the individual must belong to the same domain within your organization.

7. Two options for sharing your work outside of your organization include embedding your work in a webpage (select File > Embed) or on a SharePoint site (select File > Embed in SharePoint Online).

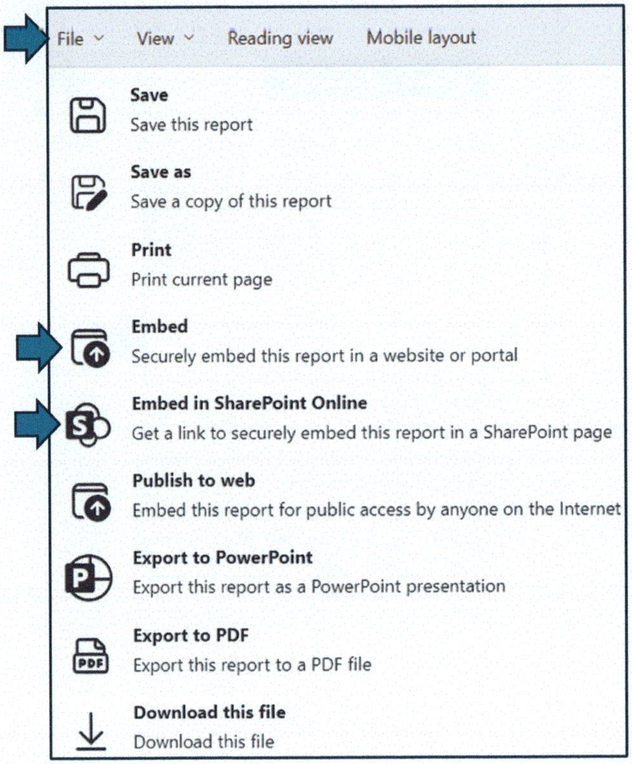

SUMMARY

In this chapter, you explored how Power BI can move from simple visualization to powerful decision-making. You learned how to make reports interactive with slicers, allowing users to focus on specific data and uncover insights tailored to their needs. You also examined Power BI's analytics tools, including forecasting, which projects future values based on historical trends. Finally, you discovered how to share your work so others can interact with your reports, interpret the data for themselves, and apply those insights across different areas of sport and business. Together, these tools show how Power BI helps analysts not only visualize data but also analyze, predict, and communicate insights effectively.

BEYOND THE BOX SCORE

1. Slicers let users filter reports by specific values. As referenced in the quote by Dante Vitagliano, how can slicers enable, inform, and improve decision-making?

2. You used Power BI to project the WNBA salary cap 5 years into the future. In what other areas of sport analytics might forecasting be useful, such as attendance, revenue, or player performance?
3. The forecast included a confidence interval. Why is it important to include uncertainty in predictions, and how can confidence intervals guide decision-making?
4. Sharing a Power BI dashboard allows others to interact with your work. Why is it important for stakeholders, such as coaches, administrators, or executives, to explore data on their own rather than only viewing static reports?
5. Now that you've completed your report and learned how to share it, how do you see yourself using dashboards in your future career or academic work, and what steps would you take to ensure your reports are clear, accurate, and impactful?

Mini Project

1. Open your Power BI dashboard from this chapter that includes player offensive and defensive statistics.
2. Use the slicer in your Power BI report to view and compare the following WNBA players:
    - Jewell Loyd
    - Kelsey Plum
    - Sabrina Ionescu
3. Which player had the highest scoring average (PTS)?
4. Which player showed the strongest defensive performance based on defensive rebounds (DREB) and steals (STL)?
5. Compare their three-point percentage (3 P%) and field-goal percentage (FG%). What do these tell you about each player's shooting efficiency?
6. How can using a slicer help a coach or analyst quickly identify player strengths and weaknesses?
7. What additional data would help give a more complete picture of players' contributions?

AI DISCLOSURE

During the preparation of this work, the author(s) used ChatGPT (OpenAI GPT-5) in order to provide editing and improve the readability of the document. After using this tool/service, the author(s) reviewed and edited the content as needed and take(s) full responsibility for the content of the publication.

REFERENCES

Alex, 2024. Uncover the story behind your data with dashboard drill-down and AI. *Syft Analytics (blog post)*. Available from: https://blog.syftanalytics.com/en/articles/9966231-uncover-the-story-behind-your-data-with-dashboard-drill-down-and-ai.

Spotrac, n.d. WNBA CBA & cap history. Spotrac. Available from: https://www.spotrac.com/wnba/cba.

Visualitics Team, 2023. Forecasting Power BI: Complete guide and practical examples. *Visualitics*. Available from: https://visualitics.it/forecasting-power-bi/?lang=en.

CHAPTER 12

Introducing Tableau

Key Takeaways

- Tableau, like Power BI, lets you turn data into interactive charts.
- You can connect files and use different visuals to spot patterns and trends.
- Learning Tableau strengthens your data visualization skills.

Without data-driven actionable insights, it's just not business intelligence.

– Aaron Parry (Parry, 2023)

Prior to Microsoft's entry into business intelligence with Power BI, Tableau was the industry standard. Now the second-largest provider of business intelligence software, Tableau is still a major player in this field (Panchal, 2024).

One of the challenges of learning Tableau is its 14-day trial. While students and educators can extend access to 1 year, the trial period can be a barrier for those who want to explore the software at their own pace.

To get started, register with Tableau at https://www.tableau.com/products/desktop/download. After registering, you will want to download and install Tableau Desktop. Once you've accessed Tableau, you can connect to a variety of data sources.

1. Import the file that contains WNBA player data. Under Connect > To a File, click Microsoft Excel (Fig. 12.1).

An Introduction to Applied Sport Analytics https://doi.org/10.1016/B978-0-443-49068-2.00012-4
© 2027 Elsevier Inc. All rights are reserved, including those for text and data mining, AI training, and similar technologies.

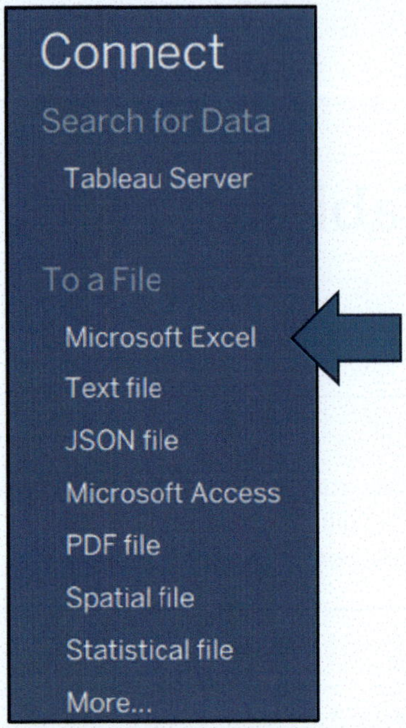

Figure 12.1 ***Selecting Microsoft Excel File.*** Process to connect Tableau with Excel file.

2. Open the file "WNBA 2024.xlsx." This file can also be downloaded from the book's website.
3. Once loaded into Tableau, our Excel file is shown to have 21 fields (statistical categories) and 36 rows (player data). Click Sheet 1 (Fig. 12.2).

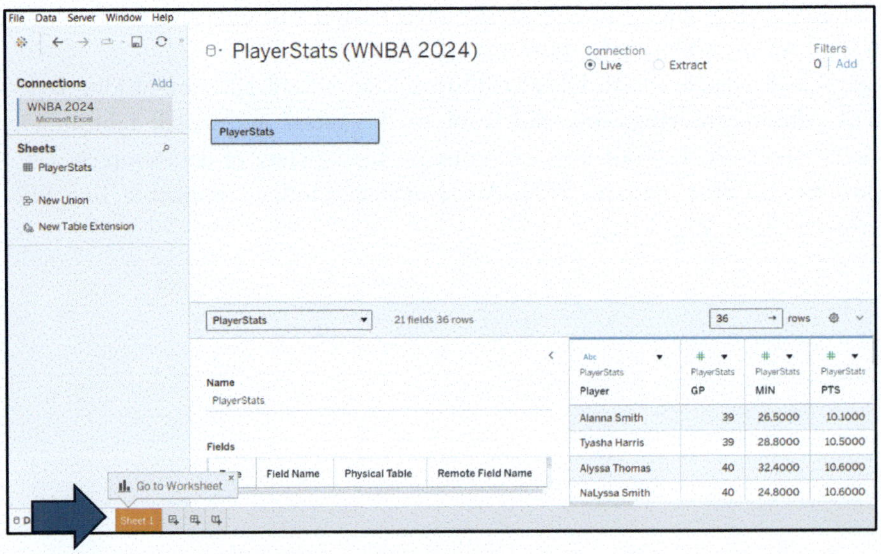

Figure 12.2 ***Excel File in Tableau.*** Sheet 1 with PlayerStats.

4. From under the Tables card, drag Player to Columns and drag 3P% to Rows. A bar chart appears (Fig. 12.3).

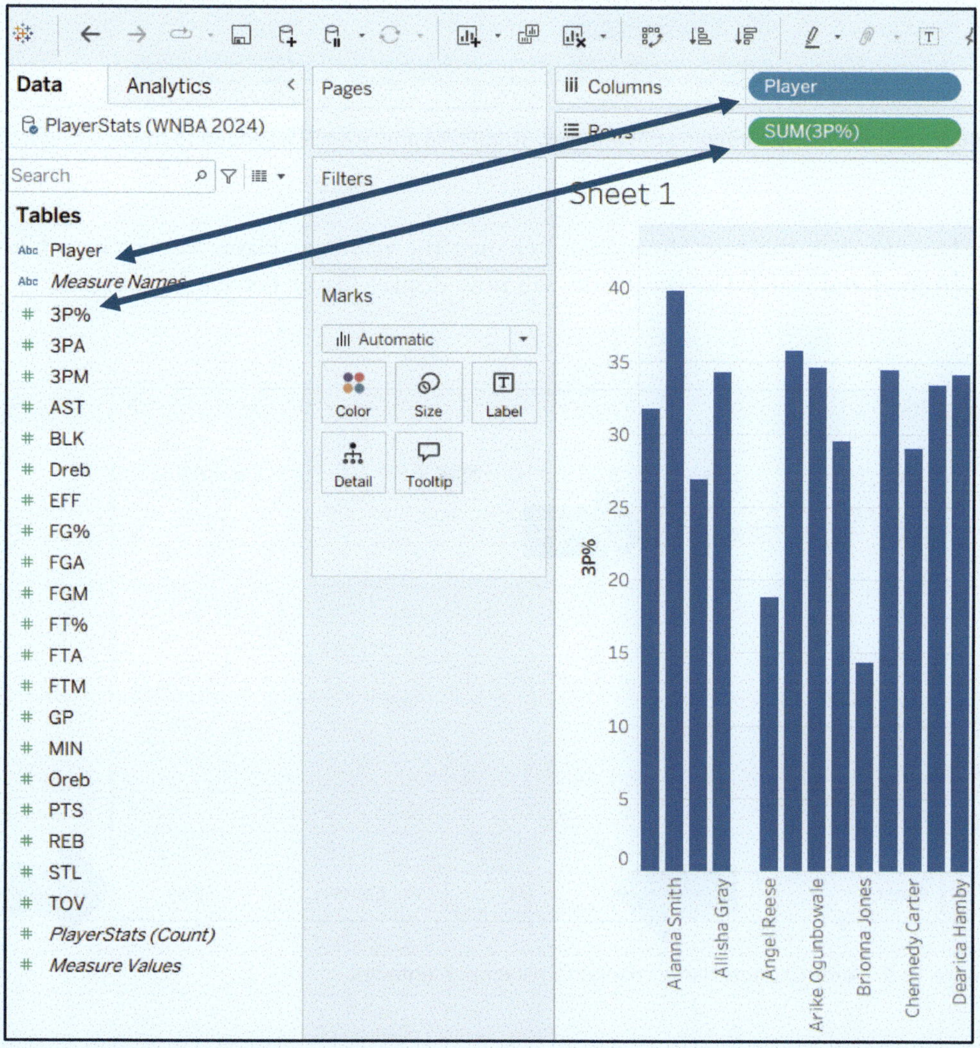

Figure 12.3 ***Dragging Variables of Interest.*** Dragging of player variable and 3P% variable.

5. Drag FG% and FT% to Rows. Three bar charts appear (Fig. 12.4).

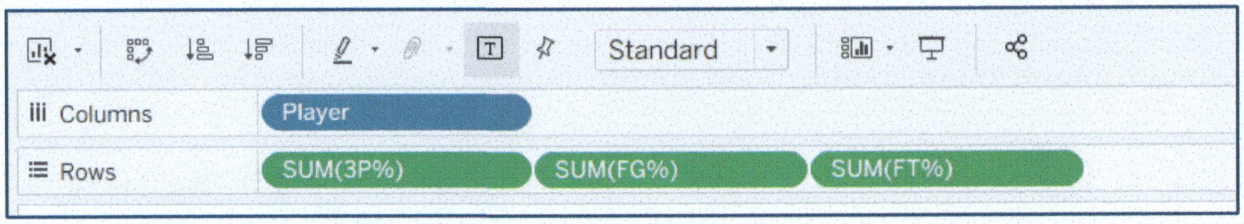

Figure 12.4 ***Additional Dragging of Variables.*** SUM(FG%) and SUM(FT%) added to rows.

6. Double-click the title Sheet 1 and change the name to Player Percentages. Click [Apply] and [OK]. The sheet title will now display Player Percentages (Fig. 12.5).

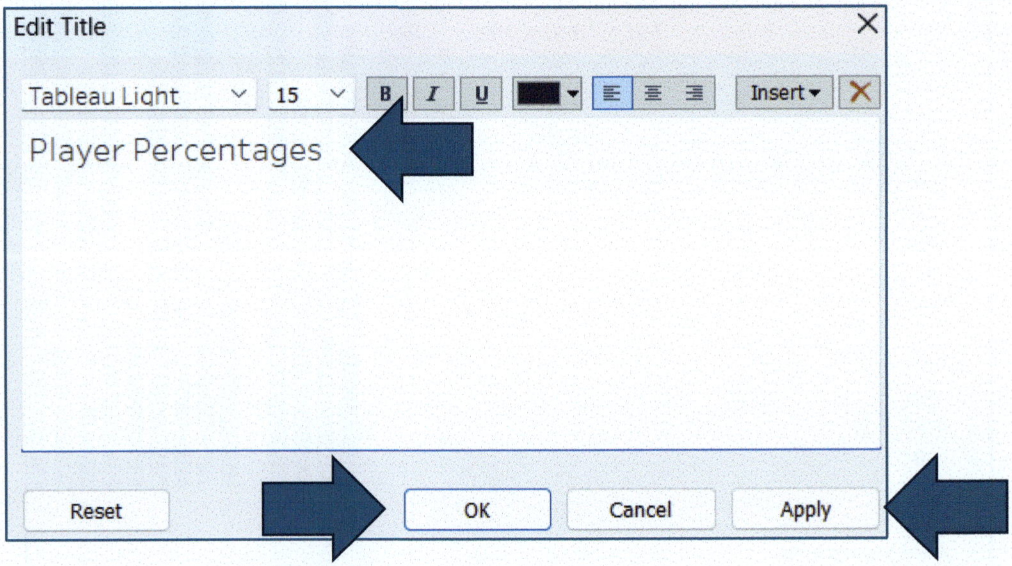

Figure 12.5 ***Name Change.*** Name change to player percentages.

7. Under the Marks card, you can adjust the color, size, labels, and tooltip content associated with the entirety of the bar chart by clicking All or you can adjust each bar chart individually (Fig. 12.6).

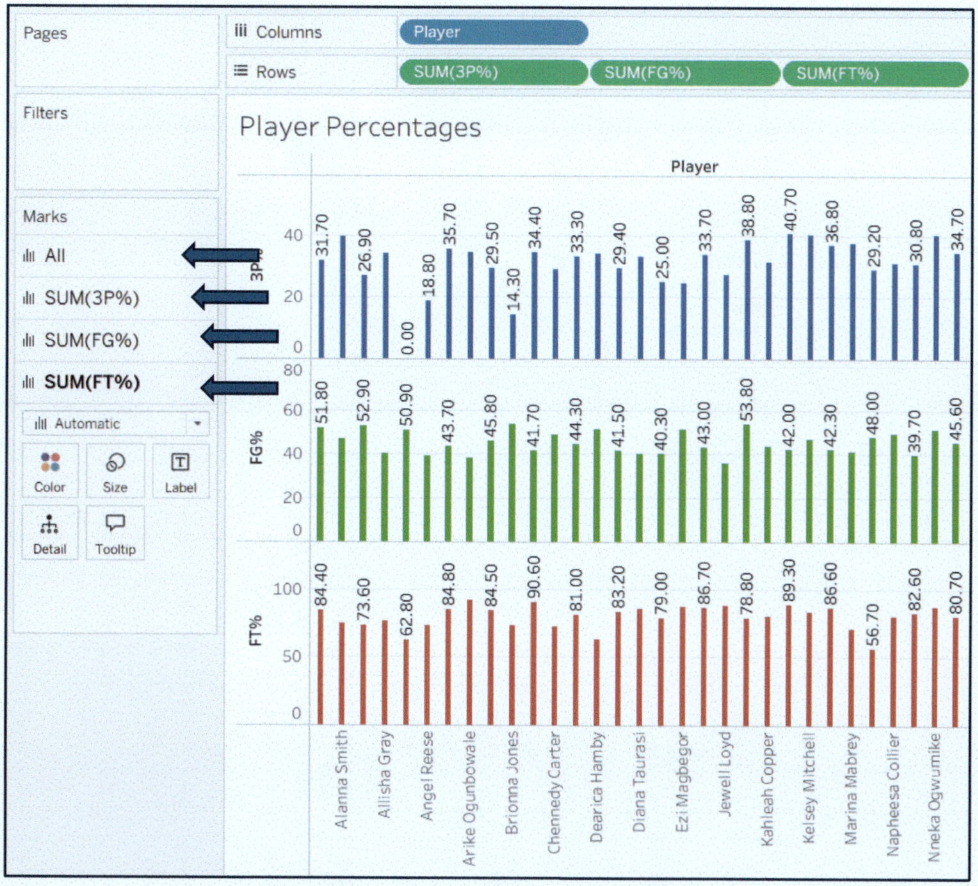

Figure 12.6 ***Three Bar Charts.*** Three bar charts of 3P%, FG%, and FT%.

8. Click the Presentation Mode icon to view the visuals as they would in a slide show. Press the escape [Esc] key to exit Presentation Mode (Fig. 12.7).

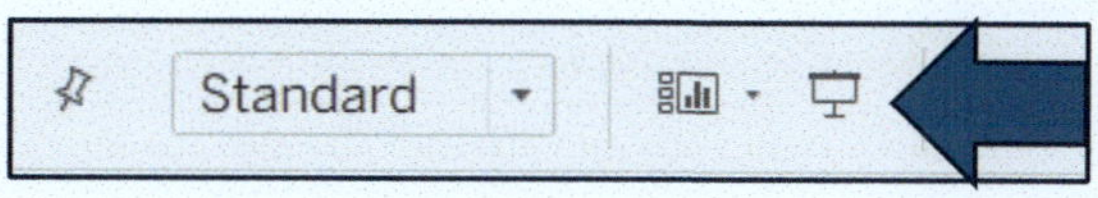

Figure 12.7 ***Presentation Mode.*** Selection of presentation mode.

A variety of chart types are available in the Show Me panel on the right side of the screen (Fig. 12.8).

Figure 12.8 ***Chart Types.*** Chart types in the Show Me panel.

Mini Project

Compare your experience using Tableau to Power BI by answering the following questions:

1. Which program was easier to connect to the dataset and why?
2. How did the process of creating visuals differ between Tableau and Power BI?
3. Which program gives you more control over formatting and design?
4. Which tool made it easier to interact with or filter your data?
5. Overall, which program would you prefer to use for sport analytics projects and why?

SUMMARY

This chapter introduced Tableau, a powerful tool for building data visualizations. The guided exercise showed how Tableau's drag-and-drop interface and formatting options help bring player performance data to life. Like Power BI, Tableau enables users to create interactive dashboards that support informed decision-making.

A NOTE ON TABLEAU

While this book includes several chapters focused on Power BI and only one on Tableau, this is an intentional choice, as Power BI is currently the most widely used business intelligence tool. Tableau remains a powerful platform, but its limited trial period makes it less practical for learning and applying data visualization techniques. That said, gaining experience with both tools can deepen your understanding of data analytics and enhance your marketability in the field.

BEYOND THE BOX SCORE

1. Aaron Parry says, "Without data-driven actionable insights, it's just not business intelligence." How does Tableau help turn raw player data into insights that could influence coaching decisions or player development?
2. In this exercise, you visualized three key shooting percentages (3P%, FG%, FT%) for WNBA players. How might comparing these metrics side by side help a coach identify player roles or make lineup decisions?
3. If your goal was to make the data "actionable," as Parry suggests, what additional information or filters would you include in the dashboard to support team strategy or front office decisions?
4. Tableau provides multiple chart types beyond bar charts. Choose one and explain how it could offer a different perspective or new insights using the same WNBA player dataset.

5. The trial limitations of Tableau may discourage long-term exploration. In your opinion, how does this affect accessibility for aspiring analysts, and what alternatives could help maintain momentum in learning data visualization tools?

AI DISCLOSURE

During the preparation of this work, the author(s) used ChatGPT (OpenAI GPT-5) in order to provide editing and improve the readability of the document. After using this tool/service, the author(s) reviewed and edited the content as needed and take(s) full responsibility for the content of the publication.

REFERENCES

Panchal, K., 2024. Power BI vs Tableau: Which is better for your business? https://www.ifourtechnolab.com/blog/power-bi-vs-tableau.

Parry, A., 2023. Unleashing business insights with Power BI: A comprehensive guide (Part I) [Blog post]. Medium. https://medium.com/learning-data/unleashing-business-insights-with-power-bi-a-comprehensive-guide-part-i-d5246adbc339.

SECTION IV

Tools of the Trade

This section serves as an introduction to some of the most important and powerful tools used in sport analytics today: SQL, Python, R, and artificial intelligence. Each chapter provides a beginner-friendly overview to help readers understand the basics of these technologies. The goal is to build familiarity and confidence with each tool, showing how they support data-driven decision-making in sport.

CHAPTER 13

SQL

Key Takeaways

- SQL helps you find, organize, and analyze data stored in databases.
- Learning SQL can give you a career advantage, since many sport and business jobs look for data skills.
- Using SQL commands like SELECT, WHERE, and JOIN lets you answer real questions, such as which players, products, or teams perform best.

Whenever you are asked if you can do a job, tell 'em, "Certainly I can!" Then get busy and find out how to do it.

– Theodore Roosevelt (Roosevelt, 2015)

SQL (pronounced "ess-que-el" or "sequel") stands for Structured Query Language. SQL is used to communicate with a database and is the most popular language for accessing, editing, and managing content in a database. This is due to its ease of use, quick processing, and reliability.

Learning SQL is important from a career perspective. At the time of writing, there are over 60,000 job listings on Indeed.com that reference SQL (Indeed, 2025). This demonstrates the demand for SQL skills across not just sports but also a variety of industries.

Because most jobs in the sport industry focus on the business side of the organization rather than on the playing field, having SQL skills can be a significant advantage on multiple fronts. Jobs in sport are often highly competitive and require graduates to start in entry-level positions. Having skills that differentiate an applicant can be considerably advantageous. Skills such as SQL can also be a key toward advancing in a career, not just for getting your foot in the door with an organization.

SQL is important for several reasons. SQL enables companies to make informed decisions based on data analysis. It helps in extracting and analyzing customer data, which can be used to improve services and products. Businesses also use SQL to understand various metrics that are crucial for their operations. Sport businesses are known to utilize SQL to look up statistics and demographics on players.

An Introduction to Applied Sport Analytics **https://doi.org/10.1016/B978-0-443-49068-2.00013-6**
© 2027 Elsevier Inc. All rights are reserved, including those for text and data mining, AI training, and similar technologies.

There are several providers that have their own variations of SQL, but which are very similar to each other, including:

- Amazon Relational Database Service
- Amazon Redshift
- Google Cloud SQL
- Microsoft Azure SQL
- MySQL—a popular open-source (free) relational database system

COMMON SQL COMMANDS

SQL includes a variety of commands for different operations. Here are some of the most common commands:

- SELECT extracts data from a database and is the most used command.
- FROM is used to specify which table to select or which data to delete from.
- WHERE is used to filter records.
- UPDATE is used to modify existing data within a database.
- DELETE is used to delete data from a database. Delete can be used to remove specific customer information, for example, or entire tables.
- JOIN combines two or more tables. This allows you to pull connected data from multiple tables into a single query.

EXAMPLES OF SQL QUERY FUNCTIONS

SQL query functions are essential for manipulating and retrieving data. Here are some examples of query functions:

1. This query requests all of the CustomerID data from the CustomerDemo table.
   ```
   SELECT CustomerID FROM CustomerDemo
   ```
2. Building upon the last example, this query requests all of the CustomerID and CustomerName data from the CustomerDemo table.
   ```
   SELECT CustomerID, CustomerName
   FROM CustomerDemo
   ```
3. This query requests all of the CustomerID and CustomerName data from the CustomerDemo table where the age of the customer is 35 years old.
   ```
   SELECT CustomerID, CustomerName
   FROM CustomerDemo
   WHERE Age = 35
   ```
4. This query requests all of the OrderNumber data in the OrderInfo table and groups the data by the state.
   ```
   SELECT OrderNumber
   FROM OrderInfo
   GROUP BY State
   ```

BECOMING FAMILIAR WITH SQL

The free lessons at w3schools.com\SQL are an ideal way to become familiar with using SQL. In addition to offering free training in SQL, w3schools.com offers a certificate and bootcamp in SQL for paying customers (W3Schools, 2025). In order to practice and use SQL, one needs to be connected to a database. The free online interface at the w3schools site utilizes a Microsoft database for Northwind Traders, a fictional global foods company. From a sport perspective, it is easy to see how such a database could also be used by a sport organization in the management of concessions. A variety of tables are provided, including:

- Customers
- Categories
- Employees
- OrderDetails
- Orders
- Products
- Shippers
- Suppliers

Let's say you are the manager of concessions for a European soccer club. Any number of questions can be answered when using SQL to query these tables, such as:

- What are the names and prices of all products sold?
  SELECT ProductName, Price
  FROM Products
- Which customers are from London?
  SELECT CustomerName, City
  FROM Customers
  WHERE City = 'London'
- What countries are our customers from?
  SELECT CustomerName, Country
  FROM Customers
  ORDER BY Country
- What are the first and last names of all employees?
  SELECT FirstName, LastName
  FROM Employees
- Which products cost more than $30?
  SELECT ProductName, Price
  FROM Products
  WHERE Price >30

Once the w3schools.com lessons are completed, training in SQL can continue by downloading the open-source software MySQL at mysql.com/downloads/ (Oracle

Corporation, 2025). Numerous books, websites, and videos are available to help in the mastery of SQL and MySQL.

SUMMARY

This chapter introduced SQL, a powerful tool for accessing and analyzing data. SQL skills are valuable for professionals in both business and sport. Developing these skills can provide you with a competitive advantage in your career.

BEYOND THE BOX SCORE

1. SQL allows sport organizations to ask specific questions of their data, such as which player sold the most merchandise or which concession item performs best. How might this ability to query data change how decisions are made within a front office?
2. "Whenever you are asked if you can do a job, tell 'em, 'Certainly I can!' Then get busy and find out how to do it," Theodore Roosevelt said. What does this quote suggest about learning new tools like SQL, and how might that mindset help you grow in a competitive field like sport management?
3. In the sport industry, decisions often need to be made quickly and under pressure. How can tools like SQL support better, faster decision-making during a season?
4. The ability to connect and combine different tables using JOIN commands is powerful. How could this help someone in a sport organization see the bigger picture rather than just isolated data points?
5. After working through the examples in this chapter, what did you learn about how databases are structured, and how might that influence the kinds of questions you ask as an analyst?

Mini Project

Go to w3schools.com/sql/sql_intro.asp and click the [Try it Yourself] button to open the SQL editor. Imagine you are querying the database of your employer, a sport stadium's concessions department. Use SQL commands to answer the following questions:

1. What are the names and prices of all products sold?
2. Which customers are from Berlin?
3. What countries are our customers located in?
4. What are the first and last names of all employees?
5. Which products cost less than $25?

AI DISCLOSURE

During the preparation of this work, the author(s) used ChatGPT (OpenAI GPT-5) in order to provide editing and improve the readability of the document. After using this tool/service, the author(s) reviewed and edited the content as needed and take(s) full responsibility for the content of the publication.

REFERENCES

Indeed (2025) SQL jobs. Available at: https://www.indeed.com/q-SQL-jobs.html (accessed 17.06.25).

Oracle Corporation (2025) MySQL downloads. Available at: https://dev.mysql.com/downloads/ (accessed 17.06.25).

Roosevelt, T. (2015) Theodore Roosevelt on bravery: Lessons from the most courageous leader of the twentieth century. New York: Skyhorse Publishing.

W3Schools (2025) SQL tutorial. Available at: https://www.w3schools.com/sql/ (accessed 17.06.25).

CHAPTER 14

Python

> **Key Takeaways**
> - Python is beginner-friendly and widely used in sport analytics.
> - Libraries like Pandas, NumPy, and Matplotlib help you clean data, analyze it, and turn it into clear visuals.
> - Learning Python builds in-demand skills.

And now for something completely different.

– Monty Python (British Broadcasting Corporation, 1969–1974)

Another choice for a tool to perform sport analytics is Python, which has seen a resurgence in recent years. First released in 1991, Python is a programming language known for its relative ease of use (Python Software Foundation, n.d.-a). Python has libraries that allow analysts to manipulate data and create custom visualizations. Python can also be used to collect, clean, and scrub data with ease. It is widely used in sport analytics, data analysis, machine learning, web development, automation, and many other fields. Python's versatility makes it an excellent choice for those wanting to venture into sports analytics.

Learning Python, like SQL, is valuable for anyone interested in data science and analytics. As of the time of writing, there are over 100,000 job listings on Indeed.com that reference Python, highlighting the high demand for Python skills across various industries, including sports (Indeed, 2025). Examples of positions in sports that involve the use of Python include biomechanics analyst, fan engagement analyst, quantitative analyst, sport data engineer, and sport scientist.

In the sport industry, data analysis is increasingly becoming a critical component of decision-making. Whether it's improving team performance, enhancing fan engagement, or optimizing business operations, Python skills can give an applicant a significant advantage in this competitive field while providing a current employee with a tool to help advance their career.

Python is important for a number of reasons, including:

- A straightforward syntax makes it easy to learn and use, even for those new to programming.
- A vast array of libraries such as Pandas, NumPy, and Matplotlib, which are beneficial for data analysis and visualization (McKinney, 2022).

An Introduction to Applied Sport Analytics **https://doi.org/10.1016/B978-0-443-49068-2.00014-8**
© 2027 Elsevier Inc. All rights are reserved, including those for text and data mining, AI training, and similar technologies.

- A large and active community, which means extensive documentation, tutorials, and forums to help troubleshoot issues and to learn new skills.
- Its versatility, as Python is used for reasons beyond data analysis, including web development, automation, and artificial intelligence.

Many important libraries are available in Python, including:

- Pandas, which is used for data manipulation and analysis
- NumPy, which is used for numerical computations
- Matplotlib, which is a library for creating data visualizations
- SciPy, which is used for scientific and technical computing
- Seaborn, which is a statistical data visualization library based on Matplotlib

There are many resources available to learn Python. Online courses are available on Python with Coursera, edX, and Udemy offering training for a fee. There are any number of books available, such as *Football Analytics with Python & R* by Eric Eager and Richard Erickson, that can help you familiarize yourself with Python. Then there's the official Python website (python.org) (Python Software Foundation, n.d.-b) and sites like w3schools.com, which provide excellent hands-on tutorials (W3Schools, n.d.-a) found here: https://www.w3schools.com/python/pandas/trypandas.asp?filename=demo_pandas_csv.

PYTHON CODE EXAMPLE

The w3schools.com website allows you to use Python with a dataset to analyze athletes. The dataset contains records of exercise sessions, detailing the duration (in minutes), pulse rate, maximum pulse rate, and calories burned for each athlete's session. The dataset is a sample of 168 individuals (W3Schools, n.d.-b). Let's say these 168 individuals are being evaluated by the performance staff at the NFL Combine. We want to assign a score to these athletes to determine which are in the top 10% based on the following formula:

$$\text{Score} = \text{Calories}/(\text{Duration} \times \text{Pulse})$$

Using the w3schools.com site, our Python code will go through the following steps:

1. Import the Pandas library. As a reminder, Pandas is used for data manipulation and analysis.
2. Read the CSV file into a DataFrame. This organizes the data into rows and columns.
3. Define a scoring function, which is our formula.
4. Apply the scoring function to each row in the DataFrame and create a new column titled Score.
5. Rank the athletes based on their scores.
6. Calculate the threshold for the top 10%.
7. Filter the DataFrame to get the individuals in the top 10%.
8. Sort the top 10% in descending order.
9. Print the individuals in the top 10%.

Here's the code we will use. You can see comments in the code for each of these steps, which appear after the # symbol. # represents a comment that helps guide the user (Fig. 14.1).

```
import pandas as pd

# Read the CSV file into a DataFrame
df = pd.read_csv('data.csv')

# Define a scoring function
def evaluate_athlete(row):
    # Example scoring: calories burned per minute per pulse rate
    score = (row['Calories'] / row['Duration']) / row['Pulse']
    return score

# Apply the scoring function to each row
df['Score'] = df.apply(evaluate_athlete, axis=1)

# Rank athletes based on their scores
df['Rank'] = df['Score'].rank(ascending=False)

# Calculate the top 10% threshold
top_10_percent_threshold = df['Rank'].quantile(0.1)

# Filter the DataFrame to get the top 10%
top_10_percent_df = df[df['Rank'] <= top_10_percent_threshold]

# Sort the top 10% by Score in descending order
top_10_percent_df = top_10_percent_df.sort_values(by='Score', ascending=False)

# Print the individuals in the top 10%
print(top_10_percent_df)
```

Figure 14.1 ***Python Code.*** Python code with directions in it.

Running the code gives us the top 17 athletes according to our formula. The chart below recreates the output based on the sample dataset from w3schools.com, with a subset of data displayed for illustrative purposes (Fig. 14.2).

User ID	Duration	Pulse	Maxpulse	Calories	Score	Rank
159	30	80	120	240.9	0.10038	1
160	30	85	120	250.4	0.09820	2
87	120	100	157	1000.1	0.08334	3
139	20	141	162	222.4	0.07887	4
72	90	100	127	700	0.07778	5
4	45	117	148	406	0.07711	6
70	150	97	129	1115	0.07663	7
95	20	151	168	229.4	0.07596	8

Figure 14.2 ***Python Output.*** Python output of top 8 athletes.

Any number of questions can be answered by analyzing this dataset. For example, we might want to know the following:

- What is the average pulse rate (Pulse) during the exercise sessions?
- What is the median number of calories burned (Calories) across all exercise sessions?
- What is the mode for the duration (Duration) of the exercise sessions?
- What are the highest and lowest values for the maximum pulse rate (Maxpulse) recorded during the exercise sessions?
- What is the 90th percentile of calories burned (Calories) during the exercise sessions?

Another free option for hands-on work with Python is PythonAnywhere.com (*PythonAnywhere:* PythonAnywhere, n.d.). This web-based Python environment can be accessed from any browser. *Head First Python* by Paul Barry is an accessible book to deepen your understanding of Python programming (Barry, 2016).

SUMMARY

This chapter introduced Python as a useful tool for sport analytics. Python's libraries can be used to organize, analyze, and visualize data. Its straightforward syntax and wide community support make it one of the most popular and accessible programming languages for beginners and professionals alike. Python's flexibility allows sport organizations to explore everything from player performance metrics to fan behavior and business operations.

BEYOND THE BOX SCORE

1. Python is used in many areas of sport, from biomechanics and performance tracking to fan engagement and revenue analysis. Why do you think Python has become such a popular tool in sport analytics?

2. Learning Python can feel intimidating for those without a background in programming. How can breaking down a project into small, manageable steps, like in the code example from this chapter, help make complex tasks more approachable?
3. In this chapter, you used Python to identify the top 10% of athletes based on a custom formula. How could this type of analysis be applied in real-world scenarios such as scouting, training, or recruiting?
4. The quote at the start of the chapter, "And now for something completely different," reflects how new tools like Python can open up fresh perspectives. How does learning a tool like Python change the way you think about solving problems in sport?
5. After working through this chapter, what did you learn about the potential of Python in sport analytics? What is one area of sport business or performance you would be interested in exploring further using Python?

Mini Project

Imagine you are part of a sport science team evaluating athlete performance data from the NFL Combine. Use Python to explore the dataset and calculate a simple performance score. Run the following code in the w3schools Pandas interface:

```
# Import the Pandas library for working with data tables
import pandas as pd
# Read the CSV file named "data.csv" into a DataFrame (like a spreadsheet)
df = pd.read_csv('data.csv')
# Create a new column called "Score" using this formula:
# Score = Calories / (Duration * Pulse)
df['Score'] = df['Calories'] / (df['Duration'] * df['Pulse'])
# Display the top 5 rows with the highest Score values
print(df.nlargest(5, 'Score'))
```

Answer the following questions using your results:

1. Which athlete recorded the highest overall Score, and what combination of Duration, Pulse, and Calories led to that performance?
2. If these athletes were wide receivers, how might differences in their Duration and Calories burned reflect their ability to sustain effort during long routes or high-tempo offensive drives?
3. What additional information would help you better evaluate these athletes' overall performance potential at the Combine?

AI DISCLOSURE

During the preparation of this work, the author(s) used ChatGPT (OpenAI GPT-5) in order to provide editing and improve the readability of the document. After using this tool/service, the author(s) reviewed and edited the content as needed and take(s) full responsibility for the content of the publication.

REFERENCES

Barry, P., 2016. Head First Python: A brain-friendly guide. 2nd ed. O'Reilly Media.
British Broadcasting Corporation, 1969–1974. Monty Python's Flying Circus [TV series]. BBC.
Indeed, 2025. Search results for "Python jobs". Available at: Indeed Python Jobs (accessed 17.06.25).
McKinney, W., 2022. Python for data analysis: Data wrangling with Pandas, NumPy, and IPython. 3rd ed. O'Reilly Media.
Python Software Foundation, n.d.-a. General Python FAQ. In: Python 3.13.5 documentation. Available at: Python Documentation FAQ (accessed 17.06.25).
Python Software Foundation, n.d.-b. Welcome to Python.org. Available at: Python.org (accessed 17.06.25).
PythonAnywhere, n.d. PythonAnywhere: Host, run, and code Python in the cloud. Available at: PythonAnywhere (accessed 21.06.25).
W3Schools, n.d.-a. Learn Python. Available at: W3Schools Python Tutorial (accessed 17.06.25).
W3Schools, n.d.-b. Pandas CSV tutorial. Available at: W3Schools Pandas CSV Tutorial (accessed 17.06.25).

CHAPTER 15

R

> **Key Takeaways**
> - R is a free, open-source programming language used for analyzing and visualizing data.
> - Thousands of R packages, including many designed for sport analytics, are available to users.
> - RStudio provides a user-friendly workspace for writing and sharing code.

R's approach makes programming much easier, partly because it offers a certain uniformity of access to data.

– Norman Matloff (Matloff, 2011)

R is an open-source programming language that is another valuable tool for data analysis and statistical computing (R: A Language and Environment for Statistical Computing, 2024). As an open-source platform, R is freely available for anyone to use, modify, and distribute. This accessibility has contributed to its widespread adoption in various fields.

Analysts rely on R for its powerful capabilities in handling, analyzing, and visualizing data. Whether it's for making data-driven decisions, identifying trends, or testing hypotheses, R provides the tools needed to perform sophisticated statistical analyses. The versatility of R makes it valuable across numerous industries and professions. Here are some areas where R is commonly utilized:

1. In academia, researchers and students in universities use R for data analysis in scientific studies, experiments, and academic projects.
2. Journalists leverage R to analyze and visualize data for stories, enabling them to present complex information in an understandable way to the public.
3. Financial analysts use R to model financial markets, perform risk assessments, and make investment decisions.
4. In the healthcare industry, R is used for analyzing clinical trial data, epidemiological studies, and bioinformatics.
5. Sports analysts apply R to evaluate player performance, optimize team strategies, and enhance training programs.

In the following example, we are evaluating athlete sprint times by performing a series of steps:

1. Create a list (vector) of athlete names.
2. Create a vector of sprint times (in seconds) for each athlete.

An Introduction to Applied Sport Analytics https://doi.org/10.1016/B978-0-443-49068-2.00015-X
© 2027 Elsevier Inc. All rights are reserved, including those for text and data mining, AI training, and similar technologies.

3. Combine the two vectors into a data frame (a table).
4. Print the data frame to view the athlete data.
5. Calculate the average sprint time.
6. Print the average sprint time.
7. Generate summary statistics for the data frame.
8. Create a basic plot of the sprint times.

In R, here is an example of what the code might look like to perform each of these steps. You can see a description of each step after the # symbol (Fig. 15.1).

```
# 1. Create a list (vector) of athlete names
athletes <- c("Riley", "Jordan", "Casey", "Drew")

# 2. Create a vector of sprint times (in seconds) for each athlete
sprint_times <- c(11.2, 10.9, 12.3, 11.7)

# 3. Combine the two vectors into a data frame (a table)
athlete_data <- data.frame(Name = athletes, SprintTime = sprint_times)

# 4. Print the data frame to view the athlete data
print(athlete_data)

# 5. Calculate the average sprint time
average_time <- mean(athlete_data$SprintTime)

# 6. Print the average sprint time
print(average_time)

# 7. Generate summary statistics for the data frame
summary(athlete_data)

# 8. Create a basic plot of the sprint times
plot(athlete_data$SprintTime,
     main = "Athlete Sprint Times",
     xlab = "Athlete (Index)",
     ylab = "Time in Seconds",
```

Figure 15.1 ***R Coding Example.*** R coding example showing calculations such as mean and summary statistics.

A benefit of using R is that it offers over 10,000 packages for data visualization, data manipulation, statistical modeling, and machine learning (Mount, 2021). Packages in R are collections of functions, data, and compiled code that extend the capabilities of R. They are designed to help users perform specific tasks more efficiently. Some popular packages include:

- ggplot2—used to create data visualizations
- dplyr—used for data manipulation and transformation
- Shiny—used for building interactive web applications
- tidyr—used for tidying up data to make it easier to work with

Many packages are also available for analyzing sports. Sports with R packages include baseball, basketball, football, hockey, and soccer, to name a few. For a well-developed list of sports packages available in R, visit cran.r-project.org/web/views/SportsAnalytics.html.

Learning R, like SQL and Python, can provide a competitive edge when applying for sport analytics jobs. Many employers value the ability to work with R due to its powerful data analysis capabilities. On Indeed.com, there are currently over 4000 jobs that reference R and SQL in the job description (Search results for "R and SQL jobs," 2025). Greater than 2,800 of these jobs pay over $100,000, demonstrating the demand for professionals skilled in R and SQL.

Just like with SQL and Python, you can gain proficiency with R at the website w3schools.com/R. A certificate is also available for a fee. From a sport perspective, the section titled "R Statistics" at w3schools.com/R features a database from Motor Trend Car Road Tests. This section is a great way to introduce the application of analytics to the sport of motor racing.

Questions that can be answered using the w3schools.com/R interface and the Motor Trend Car Road Test database include:

- What is the average number of forward gears (gear) of the cars in the mtcars dataset?
- What is the median miles per gallon (mpg) of the cars in the mtcars dataset?
- What is the mode for the variable hp (horsepower) in the mtcars dataset?
- What are the highest and lowest values for the variable qsec (which represents 1/4 mile time) in the mtcars dataset?
- What is the 95th percentile of the miles per gallon (mpg) of the cars in the mtcars dataset?

RSTUDIO

RStudio, just like R, is an open-source software that is available for download at https://posit.co/downloads/ (RStudio IDE: The most trusted IDE for open source data science, n.d.). RStudio is a graphical user interface (GUI) that allows users to work with R in a more user-friendly environment than just coding in R. RStudio is composed of several panes (Fig. 15.2).

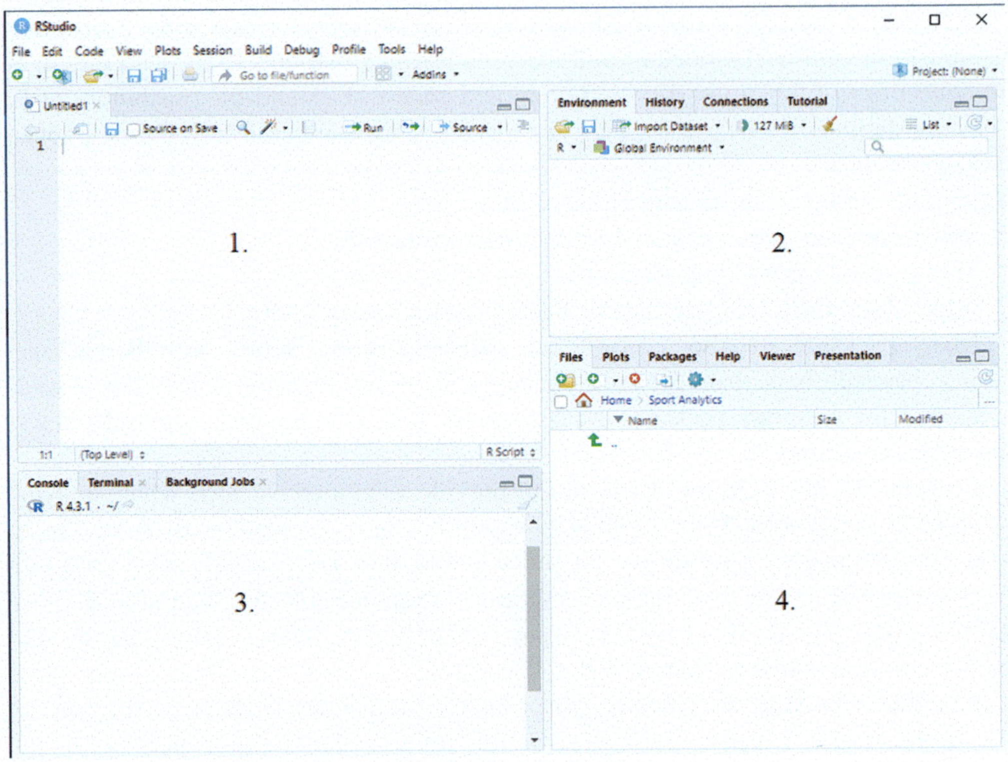

Figure 15.2 ***RStudio Blank Panes.*** Four blank panes of RStudio.

For the above image, here is a description of each pane:

1. Scripts—This is where you write and run scripts in R. A script in R is like a set of instructions and can be run repeatedly.
2. Environment—This window contains several tabs. The Environment tab lists objects specific to your current session.
3. Console—This is where the output of your script will display.
4. Miscellaneous—This window contains several tabs, but importantly, it contains the Packages tab. The packages tab provides functions, datasets, and shared code.

Let's say we want to collect and visualize team performance data for the sport of curling. We will need to create a script that extracts the necessary statistics and provides a visualization of our work. Here are the steps we will follow:

1. Load the libraries needed. These include tools for reading web pages and plotting data.
2. Create a list of team IDs. We'll use a small sample of teams for simplicity.
3. Write a function to get team data. This function grabs team name, points for (PF), and points against (PA) from a web page.
4. Create an empty table to store results. This is where the team stats will go.

5. Loop through each team and fill in the table. We get information from each web page and add it to our table.
6. Create a bar plot. Show a visual comparison of PF and PA for each team.

In the following image from RStudio, the R code is shown in the upper-left Scripts pane, and the result appears in the lower-right Miscellaneous pane under the Plots tab (Fig. 15.3).

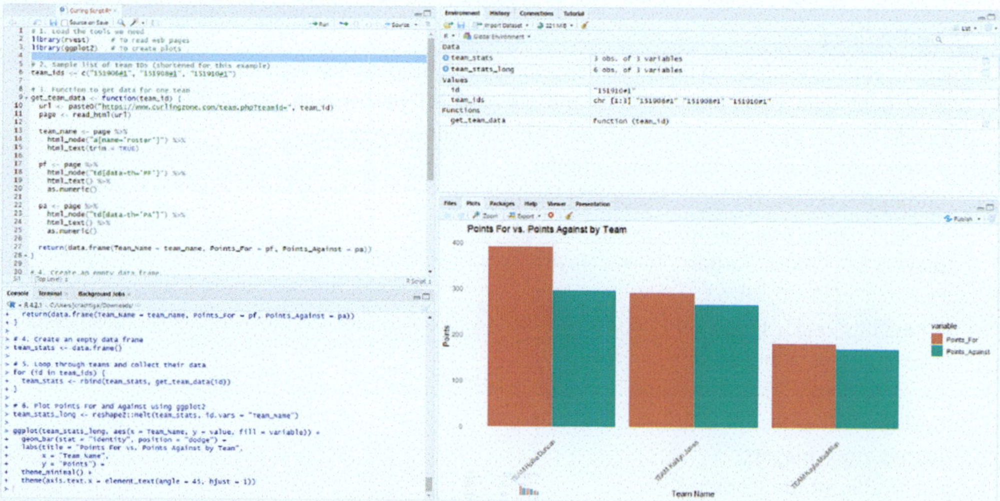

Figure 15.3 ***RStudio Filled in Panes.*** RStudio filled in panes showing coding and output.

SUMMARY

This chapter introduced R, a free, open-source programming language used for analyzing and visualizing data. R is popular in many industries because it can handle large datasets, perform statistical tests, and create data visualizations. With tools like RStudio and a wide range of packages, R is a valuable resource in sport analytics.

BEYOND THE BOX SCORE

1. R is used in a wide range of industries, from healthcare and journalism to finance and sport. What does the widespread use of R tell you about the importance of being comfortable working with data in today's job market?
2. Unlike some software that requires paid licenses, R and RStudio are free and open source. How might this accessibility benefit smaller sport organizations or individual analysts who want to explore data on their own?
3. One of R's greatest strengths is its ability to visualize and manipulate data using packages like ggplot2 and dplyr. How do these tools make it easier to communicate ideas and influence decision-making in a sport setting?

4. Imagine you work for a sport team that wants to track performance during the season. How could you use R and RStudio to help? What data would you collect, and how would you share what you found?
5. After completing this chapter, how do you see R fitting into your own analytics skill set? How might you use it in future coursework, research projects, or professional opportunities in sport?

Mini Project

Imagine you are evaluating basketball shooting performance. Use R to compare each player's overall field goal percentage (FG%) and three-point percentage (3P%). Click the [Try it Yourself] button at w3schools.com/r/ and run the following code:

```
# Needed to show graphs in the compiler
bitmap(file="out.png")
# Player names and shooting percentages
players <- c("Alex", "Jordan", "Riley", "Taylor")
FGpct <- c(55, 60, 48, 58)
ThreePct <- c(35, 15, 30, 38)
# Combine data and create bar chart
barplot(rbind(FGpct, ThreePct),
beside = TRUE,
names.arg = players,
col = c("lightblue", "orange"),
main = "FG% vs 3P% by Player",
ylab = "Percentage")
# Add legend
legend("topright", c(vFG%", v3P%"), fill = c("lightblue", "orange"))
```

Use the bar chart to answer the following questions:

1. Which player had the highest overall field goal percentage (FG%)?
2. Which player was the most accurate from three-point range (3P%)?
3. Which player seems to rely more on close-range shots rather than three-pointers?
4. What additional data might make this bar chart more meaningful?

AI DISCLOSURE

During the preparation of this work, the author(s) used ChatGPT (OpenAI GPT-5) in order to provide editing and improve the readability of the document. After using this tool/service, the author(s) reviewed and edited the content as needed and take(s) full responsibility for the content of the publication.

REFERENCES

Matloff, N. S., 2011. The art of R programming: a tour of statistical software design.
Mount, G., 2021. Advancing Into Analytics.
R: A Language and Environment for Statistical Computing, 2024. Available from: https://www.r-project.org/.
RStudio IDE: The Most Trusted IDE for Open Source Data Science. n.d. Available from: https://posit.co/products/open-source/rstudio/.
Search results for "R and SQL jobs", 2025. Available from: https://www.indeed.com/q-R-SQL-jobs.html.

CHAPTER 16

Artificial Intelligence

Key Takeaways

- Artificial intelligence (AI) can generate and refine code in R, Python, and SQL.
- AI tools like ChatGPT expand what is possible in sport analytics, from identifying player trends to scraping new data sources such as team or league websites.
- AI works best when paired with human knowledge and critical thinking.

AI is not a replacement for human intelligence; it's a tool to amplify it.

– Fei-Fei Li (Levine, 2025)

AI is a game-changer when it comes to sport analytics. While it is certainly helpful to learn how to write code in R, Python, and SQL, once you have learned the basics, AI can help elevate your work. Yes, AI such as ChatGPT can help write the code for you.

For example, let's say I want to find the top left-handed home run hitters in the history of Major League Baseball. The Lahman database is a package that is available for RStudio and contains a record of every player and at-bat in the history of professional baseball (Dalzell et al., 2024). On the site chatgpt.com, I type "Write a script for RStudio that queries the Lahman baseball package for the top left-handed hitters according to home runs all time in professional baseball." ChatGPT returns the following code (OpenAI, 2024) (Fig. 16.1).

After copying the code, I paste it into RStudio and run the code. Unfortunately, running the code in RStudio returns an error. It turns out that the Master table in the Lahman database is now called People. This change, and the schema (or table structure) for the Lahman database, is documented at https://github.com/cdalzell/Lahman/blob/master/README.md.

By making this slight change (renaming Master to People), RStudio returns the following information (Fig. 16.2).

We can see in the Console window that our script provides us with the top left-handed home run hitters all time, beginning with Barry Bonds. With AI, it's just that easy. But the question we asked was too simple. Let's investigate this issue further by asking ChatGPT to provide code that analyzes the evolution of left-handed home run hitters across different eras in Major League Baseball, accounting for factors such as

An Introduction to Applied Sport Analytics **https://doi.org/10.1016/B978-0-443-49068-2.00016-1**
© 2027 Elsevier Inc. All rights are reserved, including those for text and data mining, AI training, and similar technologies.

ChatGPT

Here's a script for RStudio that does just that:

r Copy code

```r
# Install and load the Lahman package if you haven't already
if (!require("Lahman")) {
  install.packages("Lahman")
}
library(Lahman)

# Load necessary data
batting <- Batting
players <- Master

# Merge the batting and player data to get the batting hand information
merged_data <- merge(batting, players, by = "playerID")

# Filter for left-handed hitters (Bats == "L" or Bats == "B" for switch-hi
left_handed_hitters <- subset(merged_data, bats %in% c("L", "B"))

# Aggregate the home run totals by playerID
home_run_totals <- aggregate(HR ~ playerID + nameFirst + nameLast + bats,

# Sort the data by home runs in descending order
sorted_home_runs <- home_run_totals[order(-home_run_totals$HR), ]

# Select the top 10 left-handed hitters by home runs
top_left_handed_hitters <- head(sorted_home_runs, 10)

# Display the results
print(top_left_handed_hitters)
```

Figure 16.1 ***ChatGPT Code for RStudio.*** Script for RStudio that queries the Lahman baseball package for the top left-handed hitters according to home runs all time in professional baseball.

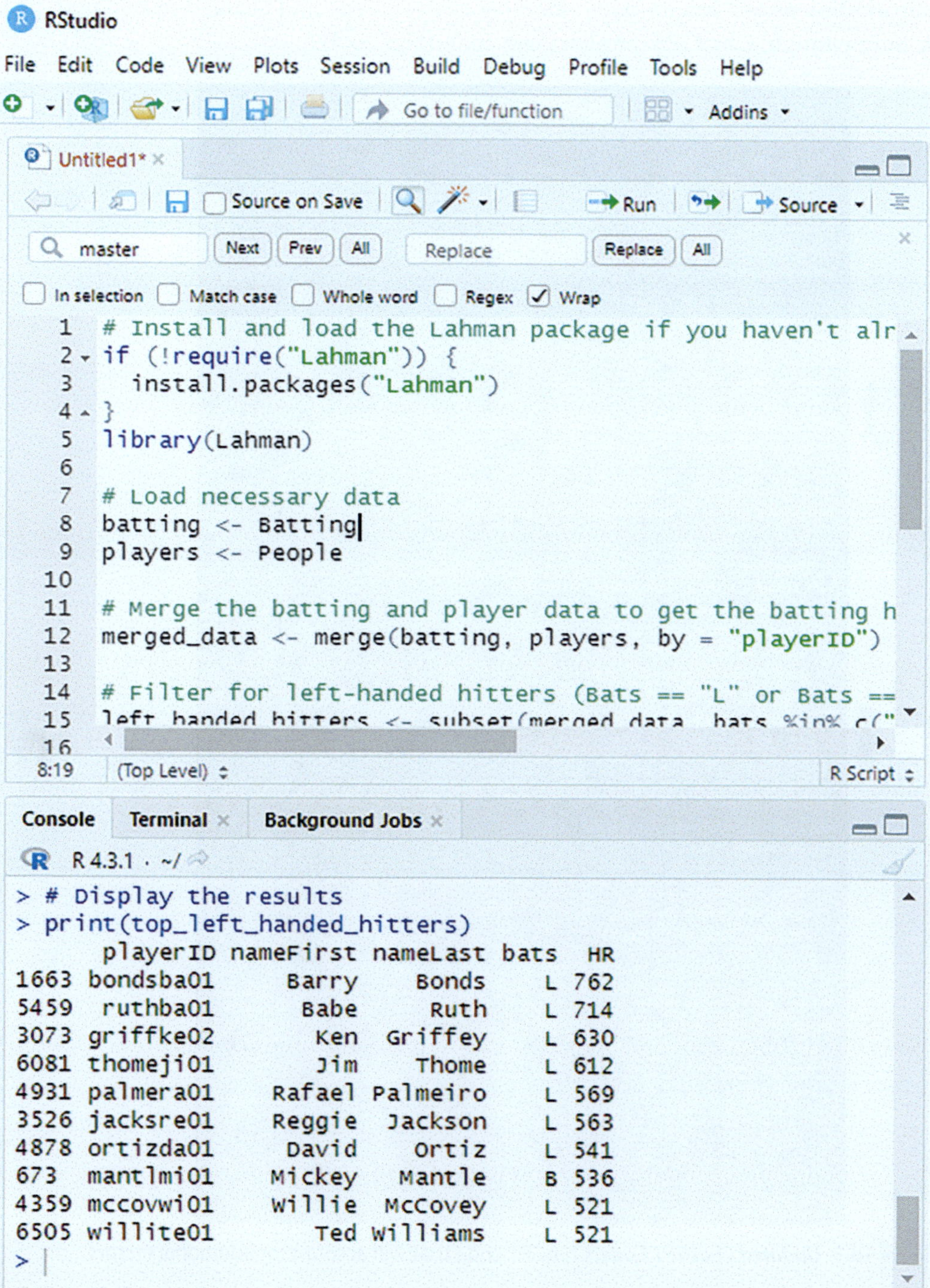

Figure 16.2 ***RStudio Results.*** RStudio results based on the last inquiry with minor corrections in the inquiry.

ballpark dimensions, league-wide offensive environment, and changes in pitching quality. Sure enough, ChatGPT provides the code (Fig. 16.3).

```r
r                                                              Copy code

# Install and load necessary packages
if (!require("Lahman")) {
  install.packages("Lahman")
}
library(Lahman)
library(dplyr)
library(ggplot2)

# Load necessary data
batting <- Batting
players <- Master
teams <- Teams

# Merge the batting and player data to get the batting hand information
merged_data <- merge(batting, players, by = "playerID")

# Filter for left-handed hitters (Bats == "L" or Bats == "B" for switch-hitters)
left_handed_hitters <- subset(merged_data, bats %in% c("L", "B"))

# Define eras
left_handed_hitters <- left_handed_hitters %>%
  mutate(era = case_when(
    yearID < 1920 ~ "Dead Ball Era",
    yearID >= 1920 & yearID < 1947 ~ "Live Ball Era",
    yearID >= 1947 & yearID < 1969 ~ "Post-War Era",
    yearID >= 1969 & yearID < 1994 ~ "Expansion Era",
    yearID >= 1994 & yearID < 2005 ~ "Steroid Era",
    TRUE ~ "Modern Era"
  ))

# Calculate home run rates (HR per AB) for each era
era_stats <- left_handed_hitters %>%
  group_by(era) %>%
  summarise(
    total_HR = sum(HR, na.rm = TRUE),
    total_AB = sum(AB, na.rm = TRUE),
    HR_rate = total_HR / total_AB
  )

# Plot home run rates by era
ggplot(era_stats, aes(x = era, y = HR_rate)) +
  geom_bar(stat = "identity", fill = "skyblue") +
  theme_minimal() +
  labs(title = "Home Run Rates of Left-Handed Hitters by Era",
       x = "Era",
       y = "Home Run Rate (HR per AB)")

# Display the results
print(era_stats)
```

Figure 16.3 ***Updated ChatGPT Code.*** ChatGPT code that will account for additional factors.

As seen below, after running the script in RStudio, we have some interesting results to consider (Fig. 16.4).

```
Console   Terminal ×   Background Jobs ×
R 4.3.1 · ~/
+   top_hitter <- home_run_totals[which.max(home_run_totals$HR), ]
+   return(top_hitter)
+ }
>
> # Find the top left-handed power hitter for each era
> top_hitters_by_era <- lapply(eras, get_top_left_handed_hitter, data = left_handed_hitters)
>
> # Combine results into a data frame
> top_hitters_df <- do.call(rbind, top_hitters_by_era)
> top_hitters_df$Era <- names(eras)
>
> # Display the results
> print(top_hitters_df)
                   playerID nameFirst nameLast bats  HR               Era
Dead-Ball Era     schulfr01     Frank  Schulte    L  92     Dead-Ball Era
Live-Ball Era      ruthba01      Babe     Ruth    L 665     Live-Ball Era
Integration Era   willite01       Ted Williams    L 430   Integration Era
Expansion Era     mccovwi01    Willie  McCovey    L 439     Expansion Era
Free Agency Era   murraed02     Eddie   Murray    B 441   Free Agency Era
Steroid Era       bondsba01     Barry    Bonds    L 486       Steroid Era
Post-Steroid Era ortizda01     David    Ortiz    L 364 Post-Steroid Era
> |
```

Figure 16.4 ***Updated RStudio Results.*** RStudio results that are based on accounting for additional factors.

We can see, according to baseball's different eras, who are the top left-handed home run hitters. These hitters include Frank Schulte, Babe Ruth, Ted Williams, Willie McCovey, Eddie Murray, Barry Bonds, and David Ortiz.

Here are the steps that the script goes through:

1. Loads and installs the Lahman package.
2. Loads the Batting and Master (People) data frames from the Lahman package.
3. Merges the batting data with the player information to get details about the players' batting hand.
4. Filters the data to include only left-handed hitters and switch-hitters.
5. Defines the start and end years for each baseball era.
6. Defines a function to get the top left-handed power hitter for a given era.
7. Applies the function to each era to find the top left-handed power hitters.
8. Combines the results into a data frame and prints the results.

But this still feels trivial. Instead, let's say you're the general manager of a team and want to know at what age left-handed batters typically peak in home run power. You can then use this information in making player personnel decisions, such as acquiring free agents or making trades. We ask ChatGPT to create an R script that looks at when left-handed batters in the poststeroid era (since 2006) peak in their home run prowess. Since we don't want just any left-handed batters, we will evaluate those players who have hit at least 50 home runs in their careers. After running the script in RStudio, we are presented with the following histogram. A histogram is a type of bar chart that groups data into frequency distributions (Fig. 16.5).

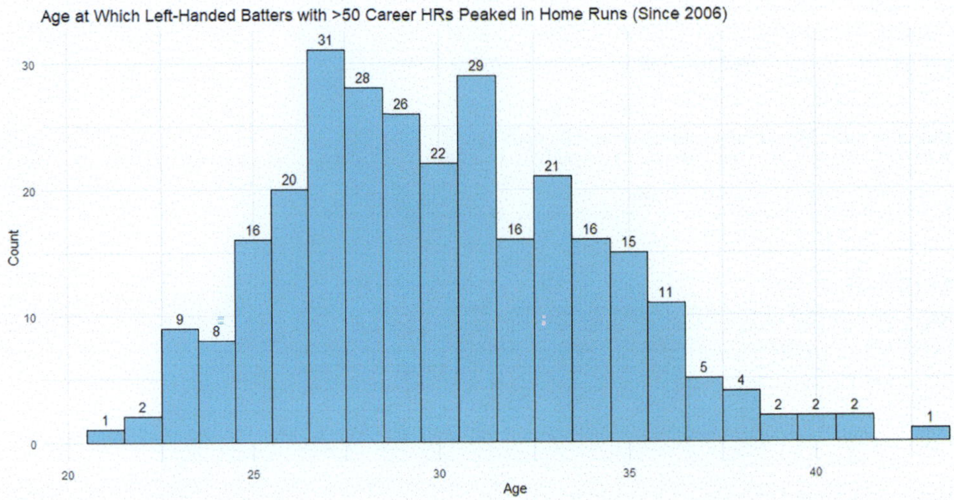

Figure 16.5 ***Histogram.*** Histogram with age as the x-axis and count of home runs on the y-axis.

Now we are getting somewhere! Here, we can see that left-handed batters are generally peaking between 27 and 31 years of age. This is critical information to you as a general manager.

Bill James once said, "I have always been much better at asking questions than knowing what the answers were" (James, n.d). With ChatGPT and a knowledge of sports and R, we are now only limited by the questions we can ask.

PYTHON AND AI

Like with R, ChatGPT and other AI can also code in Python. Let's go back to our example of using Python in the w3schools.com interface with workout information from the NFL Combine (see Chapter 14). In this example, we want ChatGPT to take our data on each athlete's exercise duration, pulse, maxpulse, and calories and view this information graphically. As a result, we ask ChatGPT for code to analyze these athletes in the form of bar chart visualizations. Our formula to score the athletes is:

$$\text{Score} = \text{Calories} \; / \; (\text{Duration} \times \text{Pulse})$$

Let's ask ChatGPT for code in Python that places these athletes into a histogram according to their scores. Here is what it provides (Fig. 16.6).

```python
# Three lines to make our compiler able to draw:
import sys
import matplotlib
matplotlib.use('Agg')

import pandas as pd
import matplotlib.pyplot as plt

# Load the data from the CSV file
df = pd.read_csv('data.csv')

# Handle missing values
df = df.dropna(subset=['Calories'])

# Calculate the score
df['Score'] = df['Calories'] / df['Duration'] * df['Pulse']


# Plot the histogram of the scores
plt.figure(figsize=(10, 6))
plt.hist(df['Score'], bins=20, color='skyblue', edgecolor='black')
plt.title('Histogram of Scores for All Athletes')
plt.xlabel('Score')
plt.ylabel('Frequency')

# Two lines to make our compiler able to draw:
plt.savefig(sys.stdout.buffer)
sys.stdout.flush()
```

Figure 16.6 ***ChatGPT Produced Code.*** ChatGPT produced code for Python.

The code works through the following steps:

1. Import the required libraries.
2. Load the data.
3. Handle any missing values.
4. Calculate the score of each athlete.
5. Plot the scores into a histogram.

Here is the histogram using the Python interface at w3schools.com for our athletes in the Combine. You can see how much easier this graphic is to interpret rather than just the data alone (Fig. 16.7).

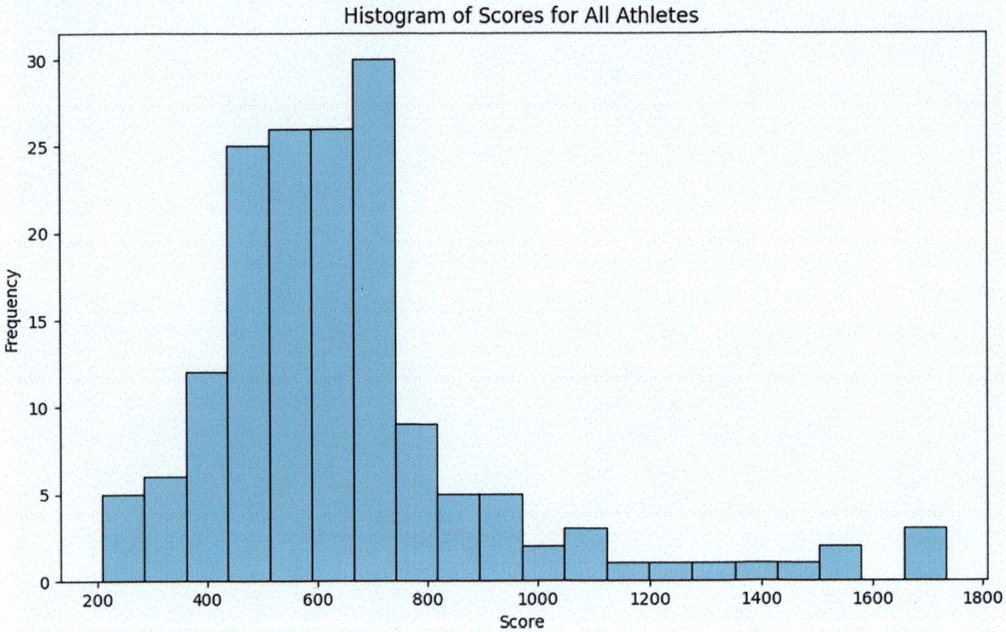

Figure 16.7 ***Histogram of Scores for Athletes.*** Histogram of scores for athletes where the x-axis is the score and the y-axis is the count.

Let's dive a little deeper into the data by asking ChatGPT for the Python code to look into the following:

"According to score (Calories / Duration × Pulse), who are the top five athletes in the combine?" Running the code provided by ChatGPT on the w3schools.com interface provides us with the following (Fig. 16.8).

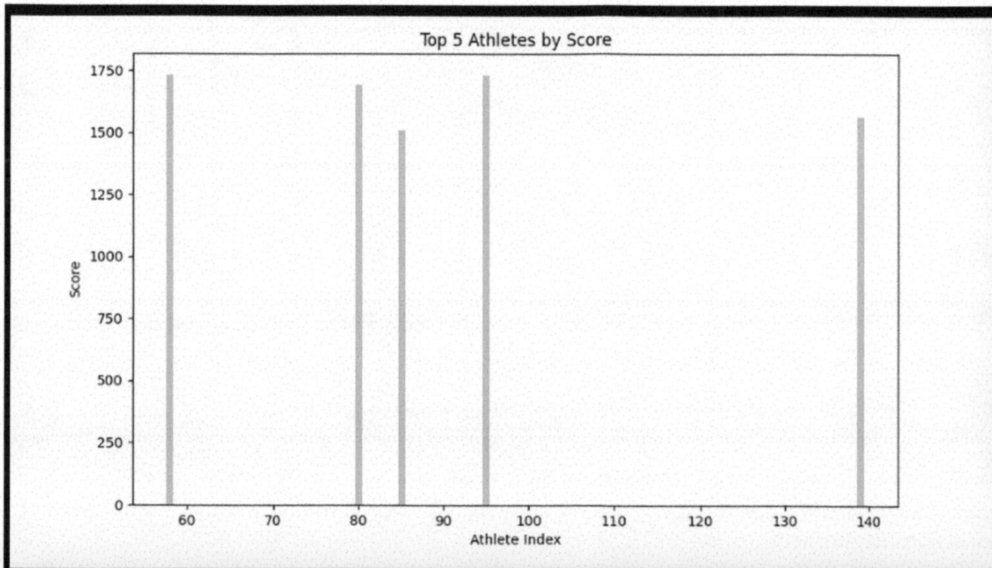

	Duration	Pulse	Maxpulse	Calories	Score
95	20	151	168	229.4	1731.970000
58	20	153	172	226.4	1731.960000
80	30	159	182	319.2	1691.760000
139	20	141	162	222.4	1567.920000
85	30	151	170	300.1	1510.503333

Figure 16.8 ***Python Results for Top 5 Athletes.*** Python results for top 5 athletes with their scores and the associated variables of calories, duration, and pulse.

By using AI's ability to code in Python, you are only limited by your data, expertise, and imagination.

SQL AND AI

Remember the w3schools.com/SQL interface uses a database that we have reimagined (see Chapter 13) for use by a concessions department of a European professional soccer team. As the manager of concessions, you want to do a deeper dive into understanding your business. As a result, you ask ChatGPT for the SQL code to identify the top 5 most productive employees. You copy the following code into the w3schools.com/SQL interface (Fig. 16.9).

```
SELECT TOP 5
    e.EmployeeID,
    e.FirstName,
    e.LastName,
    COUNT(o.OrderID) AS OrdersProcessed
FROM
    Employees AS e
LEFT JOIN
    Orders AS o ON e.EmployeeID = o.EmployeeID
GROUP BY
    e.EmployeeID, e.FirstName, e.LastName
ORDER BY
    COUNT(o.OrderID) DESC;
```

Figure 16.9 ***ChatGPT Code for SQL.*** ChatGPT code for SQL search to find the top 5 employees of a concessions department.

Here are the results when the code is run on the interface (Fig. 16.10).

Number of Records: 5

EmployeeID	FirstName	LastName	OrdersProcessed
4	Margaret	Peacock	40
3	Janet	Leverling	31
1	Nancy	Davolio	29
8	Laura	Callahan	27
2	Andrew	Fuller	20

Figure 16.10 ***SQL Output Based on Previous Inquiry.*** SQL output for the top 5 employees of a concessions department.

We can ask ChatGPT to provide us with the SQL code to analyze our customers as well. In this case, we ask for the top 5 countries with the most customers, and how many total orders have been placed by customers in those countries. We learn the following (Fig. 16.11).

Number of Records: 5

Country	TotalCustomers	TotalOrders
USA	13	29
Germany	11	25
Brazil	9	19
France	11	18
Austria	2	13

Figure 16.11 ***SQL Output Based on New Inquiry.*** SQL output for the top 5 countries with the most total orders.

Once you've learned the basics with SQL, Python, and R, AI can help make you extremely effective at data analysis.

SCRAPING WITH AI

Packages like the Lahman database in R make life easy when you want to study a sport like baseball. However, what happens when the data you need isn't available in a pre-packaged database? That is where scraping data from websites comes in.

Let's use AI and R for an example of how to scrape a website. While some knowledge of HTML can help when you run into issues, scraping with AI is often straightforward. Let's say we want to study the sport of curling, but there isn't a database available. We want to apply the Pythagorean Theorem to curling. This means we will need each curling team's won/loss record, points for, and points against. The problem is that this information isn't available on a single webpage. Instead, this information is specific to each individual team's page. This means we need to get each team's specific URL. To do this, we right click on the data and choose "View page source" (Fig. 16.12).

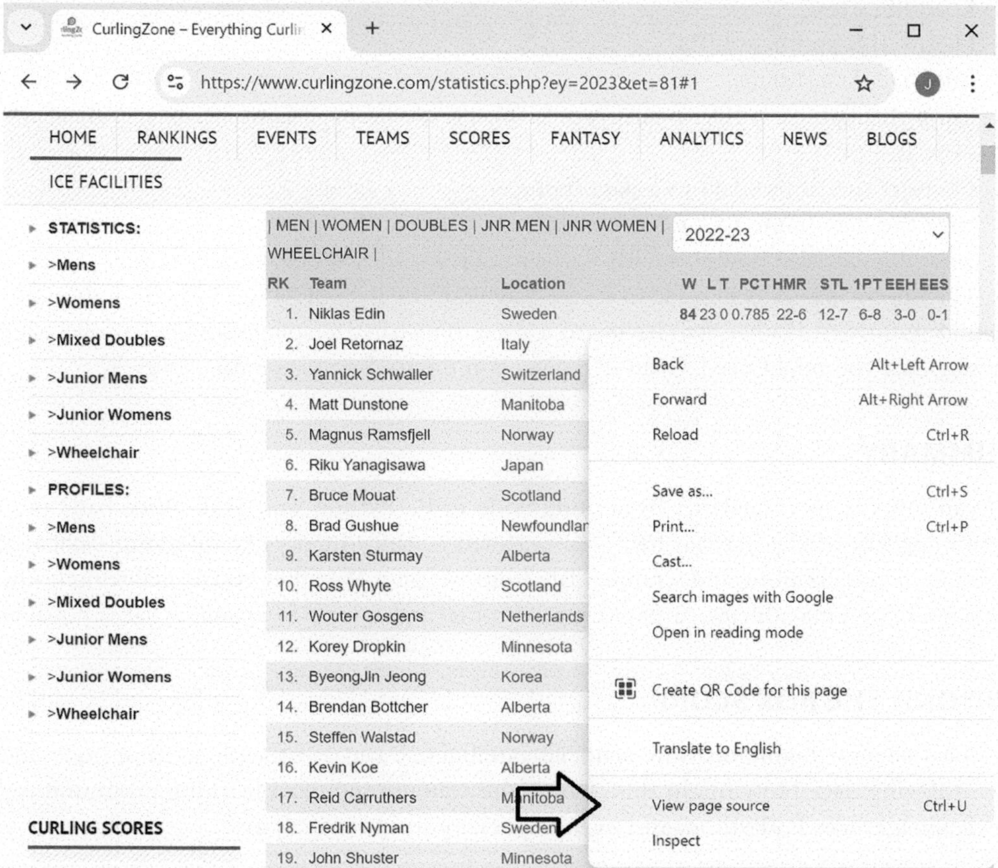

Figure 16.12 ***View Page Source Button in Curling Zone Website.*** Picture with arrow to demonstrate visually where "View page source" is in Curling Zone website.

This allows us to see the HTML code for the page and locate the URL for each team. After copying and pasting the code into Excel, we edit the spreadsheet so that we have a list of URLs for each team.

We begin by asking ChatGPT for the R code to scrape the curling website CurlingZone.com (Team stats, n.d.). Specifically, we ask:

"I would like an R Studio script that scrapes the website https://www.curlingzone.com/. Specifically, I'm looking for each team's Name, Record, PF/G, and PA/G. I'd like this scraped into an Excel spreadsheet format. Here's a sample of where the data can be found:

https://www.curlingzone.com/team.php?teamid=161080#1

Here's a list of the URLs for teams I'd like scraped from curlingzone.com:

160725#1>Michael Brunner

160726#1>Raphael Patry

160765#1>Sandy MacEwan

160767#1>Scott Jones

160806#1>Brandon Tippin"

ChatGPT responds with a script that goes through the following steps:

1. Load libraries.
2. Define IDs and base URL.
3. Initialize data storage.
4. Loop through each ID to construct URLs, read webpage content, extract data, and store it.
5. Convert the stored data to a data frame.
6. Save the data frame to a CSV file.

Scraping can also be done with Python, Excel, and Power BI.

The importance of being able to scrape data with the help of AI like ChatGPT shouldn't be underestimated. Rather than having to learn a third-party scraping software to collect any missing data you might need, ChatGPT gives you the tools to scrape data yourself.

SUMMARY

AI can support sport analytics by writing code in R, Python, and SQL. With basic programming knowledge, AI can help you perform advanced tasks like querying databases, generating visualizations, and scraping data from websites. Whether you're analyzing home run trends, ranking athletes, or exploring customer behavior, AI expands what is possible.

BEYOND THE BOX SCORE

1. This chapter highlights how artificial intelligence can assist with writing code and analyzing data. How might this change the way we approach learning programming languages like R, Python, and SQL, especially in sport analytics?

2. While AI tools like ChatGPT can generate code quickly, the output still relies on the quality of the questions we ask. How can sport analysts improve the types of questions they pose, and why does that matter?
3. As shown in the example involving left-handed power hitters, AI can help uncover insights across different eras and player profiles. How could similar techniques be applied to better understand performance trends in other sports or areas of sport business?

Mini Project

Imagine you are a sport analyst exploring how AI can assist with coding in different programming languages. Go to chat.openai.com or another AI platform and follow these steps:

1. Ask your first question and request SQL code (e.g., "Write SQL code that finds the top five players in a soccer league by total goals scored").
2. Ask your second question and request Python code (e.g., "Write Python code that creates a bar chart showing the number of points scored by five basketball players").
3. Ask your third question and request R code (e.g., "Write R code that calculates and plots the average batting average by team in baseball").

You do not need to run the code. Instead, focus on reviewing and comparing how AI responds in each language. Answer the following questions using your results:

1. Did the AI provide accurate, well-structured code for each language?
2. Which language's output seemed the easiest to understand or modify?
3. What are the benefits and potential risks of relying on AI tools to generate sport analytics code?
4. How might learning to write your own code first improve the way you use AI in sport analytics?

4. Sharing and visualizing data, such as the histograms created in R and Python, makes information more accessible and easier to interpret. Why is this important for general managers, coaches, or executives in sport?
5. After reading this chapter, what do you believe are the limits of using AI in sport analytics, and where do you see its greatest potential? How can professionals strike a balance between automation and expertise?

AI DISCLOSURE

During the preparation of this work, the author(s) used ChatGPT (OpenAI GPT-5) in order to provide editing and improve the readability of the document. After using this tool/service, the author(s) reviewed and edited the content as needed and take(s) full responsibility for the content of the publication.

REFERENCES

Dalzell, C., Friendly, M., Murphy, D., Monkman, M. and Lahman, S., 2024. Lahman: Sean 'Lahman' baseball database (Version 12.0-0) [PDF]. Available from: https://cran.rproject.org/web/packages/Lahman/Lahman.pdf.

James, B., n.d. I have always been much better at asking questions than knowing what the answers were. AZQuotes. Available from: https://www.azquotes.com/quote/995595.

Levine, A., 2025. Tales from the compliance crypt: AI is the latest specter to haunt the world of cybersecurity. *Thoropass*. Available from: https://thoropass.com/blog/compliance/tales-from-the-compliance-crypt-ai-is-the-latest-specter-to-haunt-the-world-of-cybersecurity/.

OpenAI, 2024. ChatGPT (May 13 version) [Large language model]. Available from: https://chat.openai.com/.

Team stats, n.d. Available from: https://www.curlingzone.com/.

SECTION V

Sharing Your Insights

This final section encourages readers to move from learning to contributing. It offers guidance on how to choose a topic, locate data, perform analysis, and publish your findings. The goal is not just to consume knowledge but to share it. To make a name for yourself, turn your curiosity into meaningful, public-facing work that adds to the field of sport analytics.

CHAPTER 17

Next Steps

> **Key Takeaways**
> - Publishing your research helps you find your voice as an analyst, demonstrates credibility, and contributes new insights to the sport analytics community.
> - Continual practice is the best way to deepen your understanding and confidence as an analyst.
> - Attending or presenting at conferences connects you with mentors, peers, and opportunities that can expand your career in sport analytics.

Always remember that the future comes one day at a time.

– Anonymous (Tully, 1950)

Once you start getting the hang of the ideas, tools, and coding skills in this book (and remember, no one becomes an expert overnight), the best way to keep learning is by using them. Try things out. Explore. Real growth happens when you apply what you've learned. And most importantly, share your work! Publish your research so others can see it. That's how you get better, by having an open mind to feedback. Keep moving forward in your sport analytics journey.

To start, select a sport (baseball, basketball, hockey, etc.) or an area of expertise (finance, marketing, sales, etc.) that you're comfortable with. Then, as the subject matter expert that you are, apply the principles of this book to the topic of your choosing. Remember, it is about coming up with a good question that has, to your knowledge, not been answered. This might be something that you hear an announcer say that you can investigate, or you can read other people's work, as this often leads to questions left unanswered. Locate the data you will need and then perform your analysis. Write a research article that others will find interesting, defending your conclusions with facts rather than opinions. Be sure to include data visualizations (charts, histograms, infographics, etc.) that are relevant to your research.

An Introduction to Applied Sport Analytics **https://doi.org/10.1016/B978-0-443-49068-2.00017-3**
© 2027 Elsevier Inc. All rights are reserved, including those for text and data mining, AI training, and similar technologies.

There are a number of websites where you can submit your research. For baseball, there's fangraphs.com (FanGraphs, n.d.) or any of the SABR sites at sabr.org/the-research-collection (SABR, n.d.-b). For the NFL, each team has fans who have created blogs to analyze their favorite team, or there are sites like profootballmania.com (Pro Football Mania, n.d.). If you're especially interested in football, you may want to look into the annual Big Data Bowl, an annual analytics contest that explores statistical innovations in football (Big Data Bowl, n.d.). Hockey and the NHL have versions of analytic sites as well, but it is important to note that, about a decade ago, many of the individuals who were performing hockey analysis online were hired into team offices as analysts. As a result, many of the sites offering research on the NHL have since disappeared. This type of gap might just present an opportunity for your research.

Creating your own website to publish your research is another option. Many website hosts offer either free or low-cost websites to students. I gained considerable experience with my site, FastpitchAnalytics.com, which I created as a graduate student to analyze women's college softball. The application of analytics to softball became my dissertation topic and eventually led to coaching positions in the sport. But also of importance was the chance to learn, publish my research, receive feedback, and develop connections with like-minded individuals who wanted to understand the sport through analytics.

Another great opportunity to improve your knowledge is to attend conferences. There are a number of sport analytics conferences that accept research and will allow you to present your findings. Examples include the Great Lakes Analytics in Sports Conference (Great Lakes Analytics in Sports Conference, n.d.); the Sabermetrics, Scouting, and the Science of Baseball Conference (SABR, n.d.-a); and the greatest of them all, the MIT Sloan Sports Analytics Conference (MIT Sloan Sports Analytics Conference, n.d.), to name a few. Even if you just attend one of these conferences and don't present any research, it is an opportunity to learn from and mingle with those who also have a passion for sport analytics.

And finally, never stop reading and learning. Sport analytics research is published on the web and on social media every day. There are many books that can take you beyond the start that you have received here. The keys are to apply yourself, ask questions of experts in the field, have a growth mindset where you are willing to learn, and develop your network of individuals who are willing to help. Regarding this latter point, you will be surprised how many people at the peak of sport analytics are willing to answer questions, no matter how mundane.

Creating and publishing your research is empowering and contributes to the sport analytics community. It can also help provide you with the skills needed in your career, whether those in sport analytics or otherwise. Although it can be intimidating, embrace the challenge and share your insights as you become a participant in the challenging world of sport analytics.

This final chapter serves as a capstone for everything you've learned in this book. You began by understanding the history and foundational concepts of sport analytics, from the Pythagorean Theorem and correlation to regression and expected value metrics. You then explored how to visualize and communicate insights using tools like Power BI and Tableau. You also learned to work hands-on with SQL, Python, R, and even artificial intelligence to analyze and interpret data. Together, these skills form a toolkit for answering questions, solving problems, and telling stories through sport data. As you move forward, use your technical skills, creativity, and curiosity to produce meaningful, evidence-based insights that advance the ever-growing field of sport analytics.

SUMMARY

This chapter encourages readers to apply what has been learned throughout the book by conducting and sharing their own sport analytics research. It emphasizes that growth comes from practice, feedback, and engagement with the larger analytics community. By publishing their work and connecting with others, students can turn classroom knowledge into real-world impact and professional opportunity.

BEYOND THE BOX SCORE

1. This chapter encourages you to share your work publicly. Why do you think publishing your research, even when you are still learning, is important for your growth and credibility as an analyst?
2. Thinking about your favorite sport or area of interest, what is one question that you are curious to explore using the tools and methods from this book? Why does it matter?
3. The chapter mentions the power of feedback and community. How might sharing your work with others, whether online or at a conference, help sharpen your thinking and build your confidence?
4. Visiting or presenting at a sport analytics conference is one way to engage with the field. What might you gain from attending a conference, even if you do not have a project to present?
5. "Always remember that the future comes one day at a time." How can you apply this mindset to your continued learning in sport analytics, especially when the material or tools feel challenging?

AI DISCLOSURE

During the preparation of this work, the author(s) used ChatGPT (OpenAI GPT-5) in order to provide editing and improve the readability of the document. After using this tool/service, the author(s) reviewed and edited the content as needed and take(s) full responsibility for the content of the publication.

REFERENCES

Big Data Bowl, n.d. Big Data Bowl. Available at: https://operations.nfl.com/gameday/analytics/big-data-bowl/.

Pro Football Mania, n.d. Fantasy football news and insights. Available at: https://www.profootballmania.com/.

Great Lakes Analytics in Sports Conference, n.d. Great Lakes Analytics in Sports Conference. Available at: https://www.uwsp.edu/cols-ap/sportconference.

MIT Sloan Sports Analytics Conference, n.d. MIT Sloan Sports Analytics Conference. Available at: https://www.sloansportsconference.com/.

Society for American Baseball Research (SABR), n.d.-a. Sabermetrics, Scouting, and the Science of Baseball Conference. Available at: https://sabr.org/saberseminar.

Society for American Baseball Research (SABR), n.d.-b. The Research Collection. Available at: https://sabr.org/the-research-collection/.

Tully, A., 1950. Acheson unworried by attack. Available at: https://quoteinvestigator.com/2024/03/14/one-day/.

FanGraphs, n.d. FanGraphs. Available at: https://www.fangraphs.com/.

Index

Note: Page numbers followed by "*f*" indicate figures.

R

S

Printed and bound by CPI Group (UK) Ltd, Croydon, CR0 4YY
13/09/2026
02216141-0001